2020 Austrochip Workshop on Microelectronics (Austrochip 2020)

Vienna, Austria
7 October 2020

IEEE Catalog Number: CFP20AUS-POD
ISBN: 978-1-7281-8494-4

**Copyright © 2020 by the Institute of Electrical and Electronics Engineers, Inc.
All Rights Reserved**

Copyright and Reprint Permissions: Abstracting is permitted with credit to the source. Libraries are permitted to photocopy beyond the limit of U.S. copyright law for private use of patrons those articles in this volume that carry a code at the bottom of the first page, provided the per-copy fee indicated in the code is paid through Copyright Clearance Center, 222 Rosewood Drive, Danvers, MA 01923.

For other copying, reprint or republication permission, write to IEEE Copyrights Manager, IEEE Service Center, 445 Hoes Lane, Piscataway, NJ 08854. All rights reserved.

*** *This is a print representation of what appears in the IEEE Digital Library. Some format issues inherent in the e-media version may also appear in this print version.*

IEEE Catalog Number: CFP20AUS-POD
ISBN (Print-On-Demand):   978-1-7281-8494-4
ISBN (Online):   978-1-7281-8493-7

Additional Copies of This Publication Are Available From:

Curran Associates, Inc
57 Morehouse Lane
Red Hook, NY 12571 USA
Phone: (845) 758-0400
Fax: (845) 758-2633
E-mail: curran@proceedings.com
Web: www.proceedings.com

2020 Austrochip Workshop on Microelectronics (Austrochip 2020)

Vienna, Austria
7 October 2020

IEEE Catalog Number: CFP20AUS-POD
ISBN: 978-1-7281-8494-4

TECHNISCHE
UNIVERSITÄT
WIEN

Proceedings

2020 Austrochip
Workshop on Microelectronics

978-1-7281-8494-4/20 $31.00 © 2020 IEEE

Technical Inquires:
Technische Universität Wien (TUW)
Institute of Electrodynamics, Microwave and Circuit Engineering

Contact person	Kerstin Schneider-Hornstein
Phone	+43 1 58801 354 610
Email	Kerstin.Schneider-Hornstein@tuwien.ac.at

Organisation:
Kerstin Schneider-Hornstein, Conference Chair
Filip Milicevic, Administrative Assistant
Flora Höfler, Administrative Assistant

CONTENT

Foreword	4
Sponsors	5
Steering Committee	6
Program committee	6
Invited Talk	7

Original Papers

Stefan Schmickl, Tim Schumacher, Patrick Fath, Thomas Faseth and Harald Pretl:
"A 350-nW Low-Noise Amplifier With Reduced Flicker-Noise for Bio-Signal Acquisition"   9

Tim Schumacher, Patrick Fath, Stefan Schmickl, Thomas Faseth, Philip Hetterle, Robert Weigel and Harald Pretl:
"A Design-for-Sensitivity Strategy for Charge-Pump-Based Receivers"   13

Florian Huemer and Andreas Steininger:
"Sorting Network based Full Adders for QDI Circuits"   21

Florian Huemer, Robert Najvirt and Andreas Steininger:
"Identification and Confinement of Fault Sensitivity Windows in QDI Logic"   29

Mudasir Bashir, Fatemeh Abbassi, Mirjana Videnovic-Misic, Johannes Sturm and Gernot Hueber:
"Performance Comparison of BAG and Custom Generated Analog Layout for Single-Tail Dynamic Comparator"   37

Matthias Eberlein and Harald Pretl:
"Recent Developments in Bandgap References for Nanometer CMOS Technologies"   42

Dominik Zupan and Bernd Deutschmann:
"Comparison of EMI Improved Differential Input Pair Structures Within an Integrated Folded Cascode Operational Transconductance Amplifier"   47

Martin Jungwirth, Alija Dervić and Horst Zimmermann:
"Integrated High Voltage Active Quenching Circuit in 150nm CMOS Technology"   53

Syed Sharfuddin Ahmed and Hermann Schumacher:
"Low Power Ku- and Ka-Band SiGe HBT Low-Noise Amplifiers"   57

Kerstin Schneider-Hornstein, Michael Hofbauer, Bernhard Steindl and Horst Zimmermann:
"Fully Integrated Actively Quenched SPAD in 0.18µm CMOS Technology"   62

Baset Mesgari, Horst Zimmermann, Nemanja Vokić, Bernhard Goll, Bernhard Pichler, Dinka Milovancev, Kerstin Schneider-Hornstein and Holger Arthaber:
"38.5 Gb/s RoF Based Optical Receiver for 5G Mobile Remote Radio Head Applications"   66

Sina Mortezazadeh Mahani, David Seebacher, Matteo Bassi and Johannes Sturm:
"Comparison of Bandwidth Extension Methods for Doherty Power Amplifiers for 5G"   71

Pankaj Venuturupalli, Sina Mortezazadeh Mahani, Franz Kuttner, Martin Santiago Sondón and Johannes Sturm:
"Design Procedure for a mm-Wave Transformer Based Matching Network with Spurious Rejection"   76

Markus Stadelmayer, Tim Schumacher, Thomas Faseth, Oliver Lang and Harald Pretl:
"Edge-Combining for QPSK and FSK Modulation Using A Single Delay-Line"   80

Amin Hazrati Marangalou, Ali Jalali and Masoud Meghdadi:
"An N×6M-Path Filter for Low-IF Applications: Review and Modification"   85

Amin Hazrati Marangalou, Santiago Martin Sondon, Ajinkya Kale, Johannes Sturm, Michael Ernst Gadringer and Wolfgang Bösch:
"An On-Chip Analog Spectrum Analyzer Based on Miller Frequency Divider"   90

Author Index   95

FOREWORD

The Austrochip 2020 is the 28th edition of this well-established conference. It provides a forum for exchanging ideas, presenting and discussing research results, and getting feedback back from an interdisciplinary audience consisting of physicists, engineers, and computer scientists.

This year's version is different to the normal one-day events due to the Corona pandemic. Beginning of April it became clear that the pandemic would not be over until fall, when the Austrochip traditionally takes place. The Steering Committee decided to not cancel the conference, but hold it online.

Despite the special circumstances 16 papers were chosen for presentation by the Technical Program Committee in a blind review process. The papers are published in these conference proceedings, as well as in IEEE Xplore.

I would like to thank all authors, presenters, reviewers, committee members and last but not least the sponsors of Austrochip 2020 – organizing a conference would not be possible without your valuable efforts.

I whish you all an interesting and inspiring conference and hopefully we will see each other again in person next year.

Stay safe and healthy!

Kerstin Schneider-Hornstein

Conference Chair Austrochip 2020

Institute of Electrodynamics, Microwave
and Circuit Engineering, TU Wien

SPONSORS

Technical Co-Sponsor

We would like to thank our sponsors!

STEERING COMMITTEE

Mario Huemer   (JKU Linz)
Michael Hutter   (TU Graz)
Manfred Ley   (FH Kärnten)
Timm Ostermann   (JKU Linz)
Peter Rössler   (FH Technikum Wien)
Kerstin Schneider-Hornstein (TU Wien)
Peter Söser   (TU Graz)
Andreas Steininger (TU Wien)

PROGRAM COMMITTEE

Kerstin Schneider-Hornstein (TU Wien), Conference Chair

Mario Auer (TU Graz)
Alexander Bergmann (TU Graz)
Bernd Deutschmann (TU Graz)
Dieter Draxelmayr (Infineon Technologies Austria AG)
Martin Horauer (FH Technikum Wien)
Mario Huemer (JKU Linz)
Michael Hutter (Rambus Cryptography)
Nikolaus Kerö (Oregano Systems)
Manfred Ley (FH Kärnten)
Christian Netzberger (FH Johanneum)
Burkhard Neurauter (eesy-ic)
Timm Ostermann (JKU Linz)
Eugen Pfann (JKU Linz)
Harald Pertl (JKU Linz)
Peter Rössler (FH Technikum Wien)
Gregor Schatzberger (ams AG)
Peter Söser (TU Graz)
Andreas Steininger (TU Wien)
Johannes Sturm (FH Kärnten)
Gunter Winkler (TU Graz)
Johannes Wolkerstorfer (xFace Graz)
Horst Zimmermann (TU Wien)

INVITED TALK

"Direct Time of Flight 3D Sensing and Imaging: detectors, readout circuits and system building blocks"

Speaker: David Stoppa, *ams*

Abstract:

Recently, we have assisted to a tremendous increase of the number of applications requiring highly sophisticated systems capable of taking autonomous decisions during the interaction with complex environments. Depth sensors represent a fundamental enabling technology for such applications, enabling the reconstruction of a complete 3D model of the surrounding environment, thus increasing the reliability and robustness of automatic objects classification.

Within all the possible 3D imaging technologies, Direct Time-of-Flight sensors based on Single-Photon Avalanche Detectors, SPADs, entered consumer market since a few years already, providing clear advantages in terms of distance range and ambient-light rejection. However, there are still many open challenges driving the innovation in this field at any level of the system.

This talk will provide an in-depth overview of the key building blocks for dToF systems, spanning from SPAD device technology and operation, on-chip front-end and processing electronics, up to laser module and receiving optics design parameters, through key examples from commercial and academic implementations.

Biography:

David Stoppa (SM'12-M'97) received the Laurea degree in Electronics Engineering from Politecnico of Milan, Italy, in 1998, and the Ph.D. degree in Microelectronics from the University of Trento, Italy, in 2002. In 2017 he joined AMS where he is in charge of the research and development of next generation range-sensors. From 2014 to 2017 he has been the head of the Integrated Radiation and Image Sensors research unit at FBK where he has been working as a research

978-1-7281-8494-4/20 $31.00 © 2020 IEEE

scientist since 2002 and as group leader of the Smart Optical Sensors and Interfaces group from 2010 to 2013. From 2002 to 2012 he has been teaching at the Telecommunications Engineering faculty of the University of Trento, courses of Analogue Electronics and Microelectronics. His research interests are mainly in the field of CMOS integrated circuits design, image sensors and biosensors. He has authored or co-authored more than 120 papers in international journals and presentations at international conferences, and holds several patents in the field of image sensors. Since 2011 he served as program committee member of the 'International Solid-State Circuits Conference' (ISSCC) and the SPIE 'Videometrics, Range Imaging and Applications' conference, and was technical committee member of 'International Image Sensors Workshop' (IISW) in 2009, 2013, 2015 and 2017. He was a Guest Editor for IEEE Journal of Solid-State Circuits special issues on ISSCC'14 in 2015 and he is serving as Associate Editor since 2017. Dr. Stoppa received the 2006 European Solid-State Circuits Conference Best Paper Award.

A 350-nW Low-Noise Amplifier With Reduced Flicker-Noise for Bio-Signal Acquisition

Stefan Schmickl, Tim Schumacher, Patrick Fath, Thomas Faseth, and Harald Pretl
Johannes Kepler University Linz, Austria
Email: http://iic.jku.at/analog/team/

Abstract—For the acquisition of bio-signals, like EEG or ECG, there is a need for low-noise amplifiers that are capable of amplifying input-signals with high source-impedance at very low frequencies <1 Hz, while rejecting DC-offsets >100 mV at the electrode-tissue interface, ideally without using external components. This fully differential amplifier achieves a gain of 38.7 dB with a CMRR of 74 dB and an input-referred noise of 0.97 μVrms while consuming 350 nW of power at 1 V supply voltage. This renders the amplifier well suited for RF-powered wireless bio-sensing applications. The amplifier uses a flicker-noise reducing switched capacitor feedback and a special method of metal-metal capacitor design to maximize the area-density. This fully-integrated design is implemented in a 1P6M 180 nm CMOS technology using 0.42 mm².

I. INTRODUCTION

Bio-signals acquired from the human body are very sensitive to environmental noise as the signal-amplitudes are quite low (on the order of μV), and the source-impedance is very high, easily reaching MΩ [1]. Therefore, it is beneficial to minimize the cable-lengths from the electrodes to the bio-amplifier, and in the best case the amplifier is located directly with the electrodes. To additionally overcome the annoyance of cable-bundles in multi-electrode setups, a truly wireless setup allows improving the patients comfort on the one hand and enables new possibilities of tests on the other hand. The presented bio-amplifier is the analog front-end of a monolithically-integrated battery-less wireless bio-data acquisition SoC, whose wireless front-end is described in [2].

The key requirements for the amplifier are an ultra-low power-consumption (as the targeted system is wirelessly powered), low noise and a high common-mode-rejection-ratio (CMRR). Linearity and a stable gain, on the other hand, are relaxed requirements, as the information in bio-signals is usually extracted out of relative amplitudes or rather local maximums.

Section II presents the detailed circuit implementation of the bio-amplifier including a comprehensive discussion on the influence of different feedback-variants with respect to flicker-noise performance. Section III shows measurement results of the amplifier and Section IV concludes the results.

II. CIRCUIT DESIGN

Fig. 1 shows the circuit diagram of the realized bio-amplifier, inspired by [3]. The amplifier is fully integrated and does not need any external components. The analog input

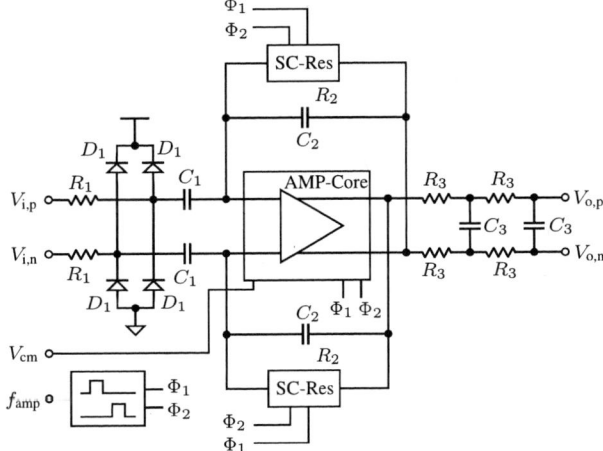

Fig. 1. Schematic of differential bio-amplifier.

terminals are directly connected to the electrodes and should, therefore, be sufficiently protected against ESD, which is realized with the series resistances R_1 and the diodes D_1 (additional ESD diodes are included with the input pads and not shown in Fig. 1). The amplifier is AC-coupled because of the potentially occurring DC-offsets at the electrode-interface due to contact voltages, which can reach up to 250 mV depending on the electrode material. The simulated input-impedance in the amplifier pass-band is 117 MΩ at 10 Hz.

Would the AC-coupling be realized as a simple RC-highpass with −3 dB-corner frequency f_c at the input, the resulting integral noise would be

$$\overline{V_{n,\text{hp}}^2} = \frac{2kT}{\pi C_{\text{hp}}}\left[\arctan\left(\frac{f_h}{f_c}\right) - \arctan\left(\frac{f_l}{f_c}\right)\right] \quad (1)$$

with k the Boltzmann constant, T the absolute temperature, the noise integration bandwidth extending from $f_l = 0.5$ Hz to $f_h = 100$ Hz, and setting $f_c = 0.1$ Hz as a compromise between signal attenuation and transient settling. With a target noise contribution of the highpass-coupling of $V_{n,\text{hp}} < 0.5\,\mu V_{\text{rms}}$, the resulting capacitance would be $C_{\text{hp}} \approx 2.1$ nF, which is much too large for chip-integration. An alternative approach is to implement an operational amplifier with capacitive impedances, as then the AC-coupling is shifted to the feedback-loop of the amplifier. The transfer function of

978-1-7281-8494-4/20 $31.00 © 2020 IEEE

an amplifier in such a capacitive-feedback-network topology (as shown in Fig. 1) is

$$A_v(s) = \frac{sC_1R_2}{1 + sC_2R_2} \qquad (2)$$

with a pass-band voltage gain of $A_v = C_1/C_2$, which is 87 as design value in this circuit and therefore should result in a maximum gain of $A_v = 38.8\,\mathrm{dB}$. The lower $-3\,\mathrm{dB}$ corner frequency can be calculated using $f_{c,l} = (2\pi R_2 C_2)^{-1}$. The capacitors C_1 and C_2 are noiseless, and the noise referred to the input of the amplifier of the large feedback resistor R_2 is given by

$$S_{n,R_2,\mathrm{in}}(f) = \frac{4kTR_2}{(2\pi f R_2 C_2)^2} = \frac{4kTR_2}{(f/f_{c,l})^2 A_v^2} \qquad (3)$$

The large thermal noise of the GΩ-sized resistor R_2 is reduced by A_v^2, and is further shaped with frequency. Thus, the effective integrated noise is largely reduced by these effects, which can ultimately be credited to the embedding of the resistor R_2 in the opamp feedback loop.

An additional advantage of using an AC-coupled amplifier in the form shown in Fig. 1 is the transfer function of the flicker noise to the output of the amplifier, which is given by

$$A_{v,\mathrm{flicker}}(s) = \frac{1 + s(C_1 + C_2)R_2}{1 + sC_2R_2}. \qquad (4)$$

Using the model $S_{n,\mathrm{flicker}}(f) = K/f$ the flicker noise at the output is thus

$$S_{n,\mathrm{flicker,out}}(f) \approx K\frac{1 + f^2(2\pi C_1 R_2)^2}{f + f^3(2\pi C_2 R_2)^2}, \qquad (5)$$

showing that for frequencies inside the passband the flicker noise is unchanged with $S_{n,\mathrm{flicker,out}}(f) = A_v^2 \cdot K/f$, but for frequencies lower than the highpass corner frequency $f_{c,l}$ the flicker noise does not increase with 1/f towards DC but rather falls with $S_{n,\mathrm{flicker,out}}(f) \approx Kf(2\pi C_1 R_2)^2$, which results in an effective reduction of the amplifier's flicker noise at the output.

Because the signals of interest already begin at low frequencies of less than 1 Hz, either R_2 or C_2 would have to be very large for such a low highpass corner, which would exceed the possibilities of monolithic integration. A common solution is to use pseudo-resistors built with MOS-transistors [4], which have the drawback of being very imprecise and underlie high process-variations. Therefore, as the highpass pole has to be quite accurate for the application, the feedback-resistor R_2 is realized as a switched-capacitor resistor (SC-res), which makes the design of comparatively accurate resistors with very large values ($>100\,\mathrm{G}\Omega$) possible. With the given component values in Table I, the lower corner frequency results in $f_{c,l} = 0.3\,\mathrm{Hz}$.

There are different ways of implementing high-value resistors as switched-capacitor circuits, two of them shown in Fig. 2. While in the shunt-type (Fig. 2(a)) [5], the charge is held on the capacitors during the cycles, and only differences are balanced, in the series-type (Fig. 2(b)) [6], the charge on the capacitors is reset to zero after every cycle. The equivalent

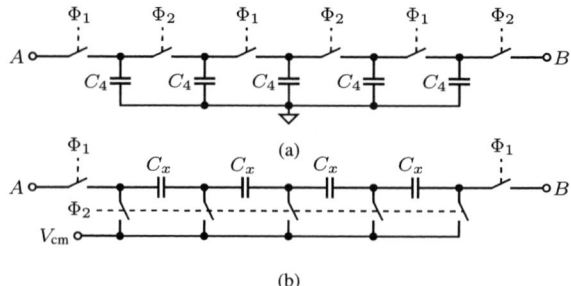

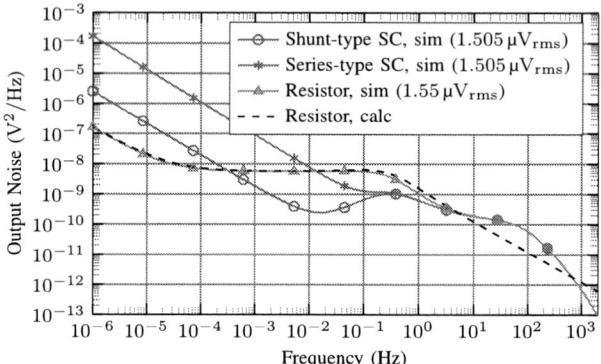

Fig. 2. Switched-capacitor resistor for amplifier feedback, (a) implemented as shunt-type and (b) the series-type.

Fig. 3. Comparison of output-referred noise of different feedback topologies. The value in parantheses notes the simulated integral input-referred noise in the bandwidth 0.5 Hz to 100 Hz.

resistors can be calculated with $R_{\mathrm{eq,shunt}} = 5/(f_{\mathrm{clk}} \cdot C_4)$ and $R_{\mathrm{eq,series}} = 4/(f_{\mathrm{clk}} \cdot C_x)$, respectively. With C_x sized accordingly for $R_{\mathrm{eq,shunt}} = R_{\mathrm{eq,series}}$, an interesting observation can be made: While the signal transfer characteristic remains the same in the simulation for both types, the flicker noise at the output is much higher using the usually-preferred series-type, as can be seen in Fig. 3. The shunt-type, in contrast, reduces the flicker-noise significantly compared to an implementation using an ideal resistor sized like the SC-implementations, also plotted in Fig. 3 for reference. The dashed line represents the calculated flicker-noise according to (5) including the thermal noise of R_2, which explains the plateau. Therefore, the shunt-type is chosen for implementation.

The SC-res, and also the common-mode control in the amplifier-core, need a non-overlapping two-phase clock-signal Φ_1 and Φ_2 [7], which are derived from a 4096 Hz clock-signal, which is also used as sample-clock of the following ADC. At the output of the amplifier-core, there is a simple 2-stage differential RC-filter with a simulated $-3\,\mathrm{dB}$ corner frequency of $f_{c,aa} = 307\,\mathrm{Hz}$. This acts as a first anti-aliasing filter stage which is planned to be followed by a biquad-filter. The simulated upper $-3\,\mathrm{dB}$ corner frequency of the complete amplifier is 87 Hz and is mainly defined by the gain-bandwidth product of the amplifier and the approx. 8 pF input capacitance of the succeeding ADC.

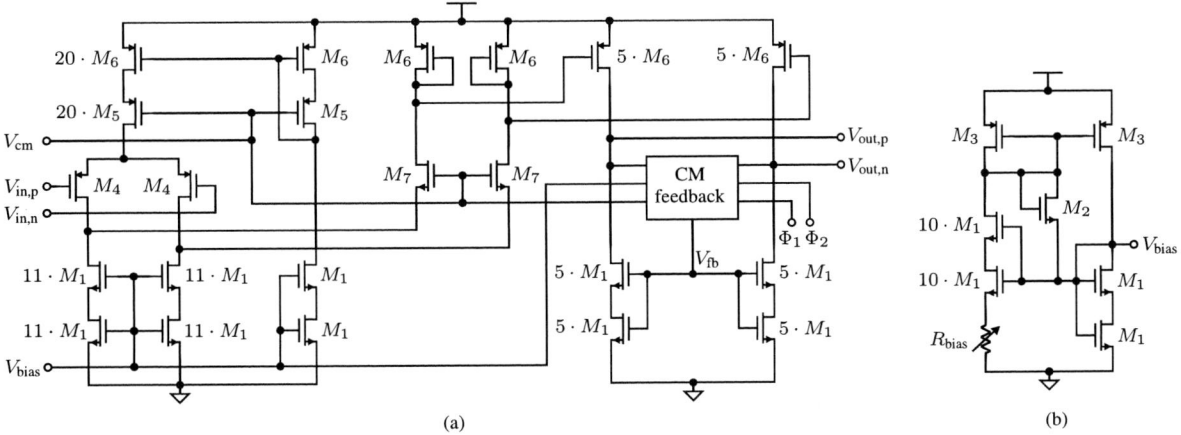

Fig. 4. (a) Schematic of amplifier core. (b) Schematic of bias-circuit for amplifier core.

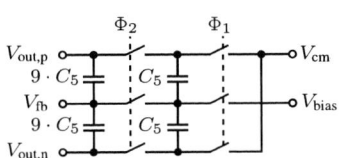

Fig. 5. Common mode feedback circuit.

TABLE I
CIRCUIT COMPONENT VALUES

Component	Value	Component	Value
R_1	108.4 kΩ	M_1	10u/15u
R_2	167.5 GΩ	M_2	5u/15u
R_3	1.08 MΩ	M_3	5u/15u
C_1	279.7 pF	M_4	250u/10u
C_2	3.2 pF	M_5	5u/10u
C_3	89.6 pF	M_6	5u/15u
C_4	6.79 fF	M_7	25u/1u
C_5	1.79 pF		

A. Amplifier Core

In Fig. 4(a), the schematic of the fully differential amplifier core is pictured. The operational amplifier consists of a folded-cascode OTA with output-stage. The bias-voltage V_{bias} is obtained from a constant-gm circuit, shown in Fig. 4(b). The current is digitally programmable in 3 steps via the resistor R_{bias}, which is implemented as 3 series resistors with two shunt NMOS to bypass them. The maximum bias-current through M_1 is 20 nA which is further used as a reference in the rest of the amplifier core. M_2 is placed in the circuit to ensure a secure startup. For a redesign, the length of M_2 should be increased to reduce the PSRR degradation. The lengths of the transistors used in the signal-chain are comparably long due to the minimization of the flicker noise and mismatch effects.

The simulated differential phase-margin of the complete amplifier (Fig. 1) is 115° and the common-mode loop gain doesn't exceed −41 dB, therefore, no further compensation measures are necessary.

The output common-mode voltage is regulated dynamically with a switched-capacitor network shown in Fig. 5 [8]. This variant of common-mode control has several well-known advantages, namely the absence of area consuming high-value resistors for the generation of the output common-mode signal and the lack of an error amplifier which would need an additional bias-current. As the loop gain with the SC-solution is much lower than with an error amplifier, it is also much easier to keep the common-mode control stable without additional compensation measures.

B. Capacitor Design

As for C_1, C_2, and C_3 large capacitances are needed, the design of a single unity capacitor was optimized. In the used technology, two capacitor types are available, namely a metal-oxide-metal (MOM) and metal-insulator-metal (MIM)-capacitor. The MIM-capacitor offers the highest density possible and is realized between the two top metal layers. This capacitor was used as a base for our unity-cell with the largest area possible according to the design rules. The area underneath was filled with a MOM-capacitor structure shown in Fig. 7(a). The two capacitor plates *Top* and *Bot* were formed by two metal lines in parallel, which are wrapped-up like in a foil capacitor. This was done on all metal layers where no MIM-capacitor could be formed. The wrapping itself does not offer a significantly larger capacitance compared to multiple parallel lines on the same area and even resulted in a slightly higher series resistance, but in combination with multiple via connections between the metal layers—forming additional capacitor plates in a vertical direction—the series resistance is reduced and a maximum of capacitance density is achieved. One benefit, compared to standard finger-MOM capacitors, is that the top-plate is completely shielded, reducing parasitic capacitance.

III. EXPERIMENTAL RESULTS

The micrograph of the amplifier, implemented in a 1P6M triple-well 180 nm CMOS technology, is shown in Fig. 7(b).

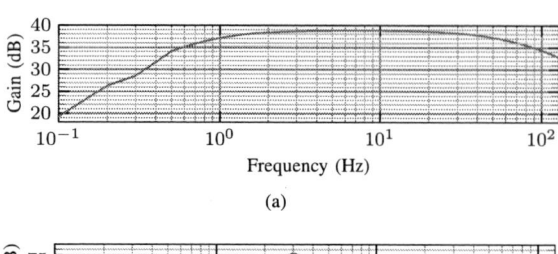

(a)

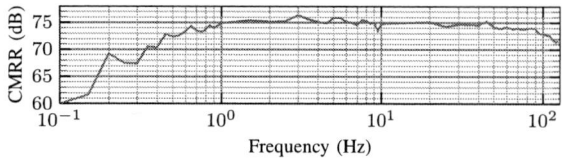

(b)

Fig. 6. Measured characteristic curves, (a) Gain and (b) CMRR.

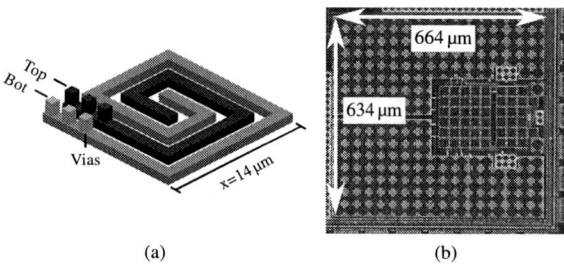

(a)   (b)

Fig. 7. (a) Visualization of realized capacitor structure, and (b) Chip micrograph.

The circuit occupies $0.421\,\text{mm}^2$. The majority of this area is filled with the input-capacitors C_1, which are implemented as unity-cells, as mentioned in Sec. II-B. The amplifier core itself is placed in the middle, also covered by MIM-caps on the top metal layers, which are also part of C_1 for best area utilization.

In Fig. 6(a) the measured gain-characteristic is shown. The peak gain is 38.7 dB, which fits very well with the calculated value in Sec. II. The lower cutoff frequency is 0.7 Hz and the higher cutoff frequency is 75.4 Hz, which are both a bit off the calculated or simulated values in Sec. II. The reason is most likely a mismatch between the fF-sized capacitors in the SC-res and the pF-sized capacitors in the capacitive feedback for the lower corner frequency, and a diverging gain-bandwidth of the amplifier for the higher corner frequency.

The measured input common-mode rejection (CMRR) is plotted in Fig. 6(b), which is > 74 dB for the considered bandwidth of 1 Hz to 100 Hz. The measured input-referred noise is $0.97\,\mu V_{\text{rms}}$, in a 0.1 Hz to 128 Hz bandwidth, calculated from the output noise by dividing through the midband-gain. The DC-current consumption of the amplifier is 350 nA at a supply voltage of 1 V. A comparison with state-of-the-art bio-signal amplifier designs is listed in Tab. II, highlighting the low power consumption and low-noise performance.

IV. CONCLUSION

This paper presents a fully integrated differential bio-signal amplifier with 38.7 dB gain, manufactured in a 180 nm CMOS

TABLE II
STATE-OF-ART BIO-AMPLIFIER COMPARISON

	[9]	[10]	[11]	[12]	This work
Process node	180 nm CMOS	180 nm CMOS	40 nm CMOS	180 nm CMOS	180 nm CMOS
V_{DD} (V)	1.35	0.2/0.8	0.6	1.2	1
Chip area (mm^2)	0.24	1	0.05	0.46	0.42
Power (nW)	18.7	790	3800	2400	**350**
Gain (dB)	36	58	30	40	38.7
R_{in} (MΩ)	93	100	140	>1000	117
CMRR (dB)	>95	85	87	>90	>74
Bandwidth (Hz)	240	670	0.5-100	0.5-500	0.7-75.4
Input-referred noise (μV_{rms})	2.45	0.93	1.45	1.8	**0.97**
PEF/NEF	0.99 / 0.86	1.6 / 2.1	119 / 14	23 / 4.4	6.6 / 2.6

$$\text{PEF} = v_{\text{ni,rms}}^2 2 P_{\text{tot}}/(V_{\text{T}}4k_{\text{B}}T\pi BW), \quad \text{NEF} = v_{\text{ni,rms}}\sqrt{2I_{\text{tot}}/(V_{\text{T}}4k_{\text{B}}T\pi BW)}$$

process. The amplifier is part of a wirelessly powered SoC for bio-signal acquisition with integrated ADC and transceiver-frontend. With a highpass corner of 0.7 Hz, the amplifier is capable of handling large DC-offset voltages while enabling amplification of signals starting at < 1 Hz which typically appear in EEG neural recordings. Because of the low active power of 350 nW, but also a very low input-referred noise of $0.97\,\mu V_{\text{rms}}$, the amplifier is well suited for battery- or RF-powered devices. The combination of extremely low power and very low noise was enabled by exploiting the advantages of switched capacitor techniques.

REFERENCES

[1] Y. M. Chi, T.-P. Jung, and G. Cauwenberghs, "Dry-Contact and Non-contact Biopotential Electrodes: Methodological Review," *IEEE Rev. Biomed. Eng.*, vol. 3, pp. 106–119, 2010.

[2] S. Schmickl, T. Faseth, and H. Pretl, "An RF-Energy Harvester and IR-UWB Transmitter for Ultra-Low-Power Battery-Less Biosensors," *IEEE Transactions on Circuits and Systems I: Regular Papers*, pp. 1–10, 2020.

[3] N. Verma, A. Shoeb, J. Bohorquez, J. Dawson, J. Guttag, and A. P. Chandrakasan, "A Micro-Power EEG Acquisition SoC With Integrated Feature Extraction Processor for a Chronic Seizure Detection System," *IEEE Journal of Solid-State Circuits*, vol. 45, no. 4, pp. 804–816, 2010.

[4] T. Delbruck and C. A. Mead, "Adaptive photoreceptor with wide dynamic range," in *Proceedings of IEEE International Symposium on Circuits and Systems - ISCAS '94*, vol. 4, 1994, pp. 339–342 vol.4.

[5] D. L. Fried, "Analog sample-data filters," *IEEE Journal of Solid-State Circuits*, vol. 7, no. 4, pp. 302–304, 1972.

[6] J. T. Caves, S. D. Rosenbaum, M. A. Copeland, and C. F. Rahim, "Sampled analog filtering using switched capacitors as resistor equivalents," *IEEE Journal of Solid-State Circuits*, vol. 12, no. 6, pp. 592–599, 1977.

[7] P. E. Allen, *Switched Capacitor Circuits*. Van Nostrand Reinhold Company, 1984.

[8] D. Senderowicz, S. F. Dreyer, J. H. Huggins, C. F. Rahim, and C. A. Laber, "A family of differential NMOS analog circuits for a PCM codec filter chip," *IEEE Journal of Solid-State Circuits*, vol. 17, no. 6, pp. 1014–1023, 1982.

[9] S. Mondal and D. A. Hall, "A 13.9-nA ECG Amplifier Achieving 0.86/0.99 NEF/PEF Using AC-Coupled OTA-Stacking," *IEEE Journal of Solid-State Circuits*, vol. 55, no. 2, pp. 414–425, 2020.

[10] F. M. Yaul and A. P. Chandrakasan, "A Noise-Efficient 36 nV/ $\sqrt{}$ Hz Chopper Amplifier Using an Inverter-Based 0.2-V Supply Input Stage," *IEEE Journal of Solid-State Circuits*, vol. 52, no. 11, pp. 3032–3042, 2017.

[11] J. Xu, Q. Lin, M. Ding, Y. Li, C. Van Hoof, W. Serdijn, and N. Van Helleputte, "A 0.6V 3.8µW ECG/bio-impedance monitoring IC for disposable health patch in 40nm CMOS," in *2018 IEEE Custom Integrated Circuits Conference (CICC)*, 2018, pp. 1–4.

[12] Q. Li, X. Wang, and Y. Liu, "A 60 nV/ $\sqrt{}$ Hz <0.01%-THD ±200 mV-DC-Rejection Bio-Sensing Chopper Amplifier With Noise-Nonlinearity-Cancelling Loop," *IEEE Transactions on Circuits and Systems II: Express Briefs*, vol. 67, no. 2, pp. 215–219, 2020.

978-1-7281-8494-4/20 $31.00 © 2020 IEEE

2020 Austrochip Workshop on Microelectronics

A Design-for-Sensitivity Strategy for Charge-Pump-Based Receivers

Tim Schumacher[*] Patrick Fath[*], Stefan Schmickl[*], Thomas Faseth[*], Philip Hetterle[†], Robert Weigel[†] and Harald Pretl[*]

[*]Johannes Kepler University Linz, Institute for Integrated Circuits, Austria
[†] Friedrich-Alexander-University Erlangen-Nuremberg, Institute for Electronics Engineering, Germany

Abstract—**This work presents the theoretical fundamentals of designing charge-pump-based receivers for maximum open-circuit voltage sensitivity (OCVS). From a classical Dickson charge pump with sub-zero-V_{th} transistors, various models are derived to extract correlation between specific parameters (e.g. number of stages, transistor widths and capacitor sizes) and the output voltage, maximum data rate and input impedance. This approach enabled us to develop an easily applied design strategy for this type of low-power receivers, which we demonstrate here for the 868 MHz and 915 MHz short-range-devices bands. It also considers an external matching network, our design strategy provides flexibility in layout generation and allows fast designing of receivers. Our strategy focuses on a rapid design flow rather than accurate representation, and is demonstrated using the example of a receiver design. In addition, the limitations of the derivations are discussed, and further optimizations are suggested.**

Index Terms—**Charge-pump, design-automation, matching, passive, receiver, SRD, wake-up.**

I. Introduction

Charge pumps are commonly used in energy-harvesting architectures and are designed for specific input levels to provide maximum efficiency [1], [2]. Hence, they are limited to short distances of only a few meters. This work focuses on an alternative approach using a charge pump: rather than for energy-harvesting, the charge-pump is optimized as an envelope-detector receiver for on-off keying (OOK) modulated signals with low input levels. A Dickson charge pump [3] offers a simple, area-efficient, and –most importantly– passive design for realizing an ultra-low-power and easy-to-implement receiver. This is especially interesting for short-range devices (SRD)-band applications and sensor nodes. However, realizing the charge pump brings with it the problem of an input impedance that depends on the input power level. To avoid this and maintain a fixed input impedance, a buffering amplifier could be used, since buffering stages and pre-amplification cannot be employed in an ultra-low-power design, the constraints for a corresponding matching network would be challenging. Further, the chargeing time of the charge pump –or in this case– envelope-detector, limits the maximum achievable data rate. Numerous parameter such as number of stages and the physical dimensions of the transistors, simultaneously affect both input resistance and charging time, which again limits the data rate.

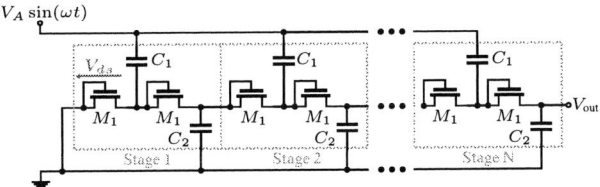

Fig. 1. Rectifier based on RF energy harvester (type A in [4]) using sub-zero–V_{th} NMOS transistors and N stages.

To address these problems, this work proposes approximates models for a rapid and easy design flow for charge-pump-based envelope-detector receivers. The modulations were carried out for sub-zero–V_{th} transistors, which maximize the OCVS, and verified by simulation in a 180 nm 1P6M technology. Based on the approximate models, a design strategy for charge-pump-based receivers is presented, that even allows a fully automated design flow to be implemented.

In Section II, the theoretical equations and models are derived and their limitations discussed. On this basis and using the proposed strategy, an example receiver front-end was designed and then realized, as described in Section II. Further, the layout of the example design is shown, and simulation results for the layout after parasitic extraction are compared to the results for the modeled receiver. Finally, Section IV concludes the paper.

II. Modeling of a Charge-Pump-Based Receiver

This section presents the theoretical model for describing charge-pump-based rectifying circuits with optimized sensitivity and sub-zero–V_{th} transistors. The term sub-zero–V_{th} emphasizes that the effective threshold voltage of these transistor lies below zero volts. Due to their leakage current, sub-zero–V_{th} transistors are avoided in classical designs, but, in the given use case as an envelope detector for data transmission at very low-input amplitudes, they offer higher sensitivity and data rates, as shown below. First, the main function of a charge-pump is recapitulated, as shown in Fig. 1. For example, a Dickson charge pump consists of a series of diode-connected FETs (M_1) with additional taps that are capacitive-coupled via C_1. Applying a sine-wave shaped signal $V_A \sin(\omega t)$ to the taps, forces the corresponding node voltages to follow

978-1-7281-8494-4/20 $31.00 © 2020 IEEE

this oscillation. If a node voltage falls below the threshold of a FET diode, a charge flow is created. As an ideal diode allows the charge to flow in only one direction, the DC node voltage is increased. For the non-differential pump shown, additional storage capacitances C_2 are needed. One tap- and one charging-capacitance in combination with their corresponding transistors are referred to as one stage of the charge pump. Since the charge pump compromises N stages, the final stage's output voltage is a multiple of the input envelope (in this simplified case V_A) minus a particular drop over each diode or transistor.

With the main function modified, the subsequent sections provide approximations and modulations so that all major design variables, such as number of stages, transistor sizing, capacitor sizes and an additional matching network for the incoming signal, can be determined rapidly. The values thus obtained require slight adaptation after layout design and parasitic extraction, but the aim of this work is to present a fast yet feasible estimation, to render excessive prior analysis or parameter sweeping unnecessary. The approximations used impose certain limitations, which are mentioned throughout the following sections and are summarized in Sec. II-G.

A. Output Voltage

First, the general output voltage is derived by using a charge-pump based on sub-zero-V_{th} transistors, as shown in Fig. 1. It is one of many variants of a classical Dickson charge pump as presented in [4] and [5]. The reasons for choosing this topology are its single-ended design, which can easily be attached to miniature ceramic chip antennas, and increases sensitivity [6], and its short charging and discharging times as it must follow the envelope of amplitude-modulated signals with typical data rates of up to 100 kHz. In [6], a similar charge-pump was used as an energy-harvester structure with −32 dBm sensitivity. Consequently, this design could also be used as a receiver, but with documented charging times of several microseconds, data rates of more than 1 kHz cannot be realized. Thus, this structure must be optimized for data reception. In [6], a steady-state approximation of the output voltage of the transistor chain was presented

$$V_{\text{out,sub-Vth}} = 2N \cdot V_T \cdot \ln\left[I_0\left(\frac{V_A}{V_T}\right)\right] \qquad (1)$$

where I_0 is the modified Bessel function and V_A the amplitude of the input signal of the charge pump. This approximation assumes that all transistors operate in the sub-threshold region and that the capacitive output is fully charged. As can be seen from the equation, the steady-state voltage depends neither on the threshold voltage nor the dimensions of the transistors used. Rather the transistor parameters affect input resistance and charging time of the charge pump. The Authors of [4], [6] showed that by using zero–V_{th} devices, the charging time can be reduced. Zero–V_{th} devices in 180 nm 1P6M technology as used here as an example offer an effective threshold V_{th} of less than −40 mV, which causes the diode-connected devices to be in the linear region for small input amplitudes, and are

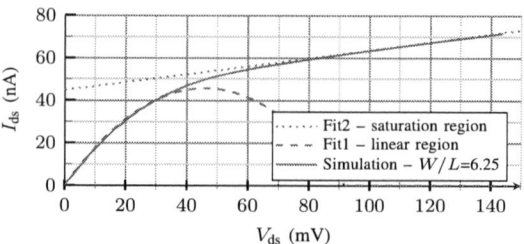

Fig. 2. Fitting of diode-connected transistor equations for linear and saturation region.

therefore referred to as sub-zero–V_{th} devices. This behavior is illustrated in Fig. 2, where a simulated drain current of a diode-connected transistor is shown. The basic transistor equations for the linear and saturation region were fitted to the simulated curve, indicating a V_{th} of −46 mV in the linear region. For simplicity, the bulk effect was ignored and a mean value chosen for V_{th}. This also causes the deviation in the transition region and consequently prevents crossing of the two fitted curves shown in Fig. 2.

For input amplitudes below $|V_{th}|$, the transistor stays in the fitted linear region and thus (1) is not applicable, and the gap between the two curves in Fig. 2 is not critical. This means that a new equation to describing the output voltage is needed, when operating the device in the linear region only. In [7] and [6], where the output voltage for mixed or saturation regions was determined, the output voltage was generally calculated by

$$V_{\text{out}} = 2N \cdot (V_A - V_{\text{drop}}), \qquad (2)$$

where each transistor offers an ideal voltage boost of V_A (the input amplitude) minus a voltage V_{drop} that represents the drop over a single diode-connected transistor. Given that in steady state, the charge transferred through a single diode-connected transistor equals zero and all sub-zero-V_{th} transistors are in the linear region, an equation for V_{drop} can be formulated

$$Q = \int_0^T I_{ds}\,dt = \int_0^T K\left((V_{\text{gs}} - V_{\text{th}})V_{\text{ds}} - \frac{V_{\text{ds}}^2}{2}\right) dt = 0, \qquad (3)$$

where $K = \mu_n C_{ox} W/L$ is a parameter that depends on the technology used. The time-dependent drain-source voltage V_{ds} shown in Fig. 1 can be replaced with

$$V_{\text{ds}} = V_A \sin(\omega t) + V_A - V_{\text{drop}}, \qquad (4)$$

where the first term on the right-hand side is the oscillation of the input signal and $V_A - V_{\text{drop}}$ plus the oscillation of the input signal. Inserting (4) into (3) yields the voltage drop V_{drop} as

$$V_{\text{drop}} = \pm\frac{1}{2}\sqrt{4V_{\text{th}}^2 - 2V_A^2} + V_{\text{th}} + V_A. \qquad (5)$$

Since V_{th} is negative for our sub-zero–V_{th} transistors, only the positive solution is physically. This allows a new equation for

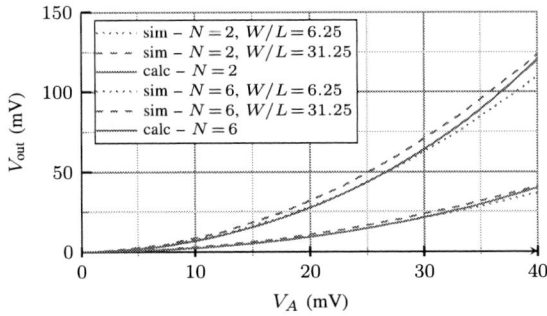

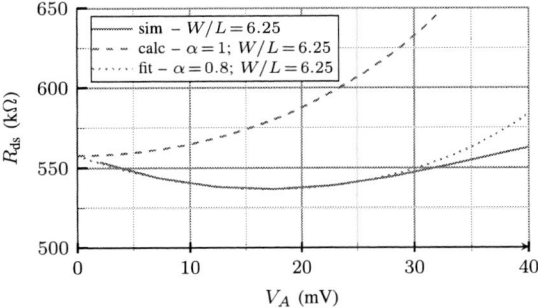

Fig. 3. Output voltage of the charge pump from Fig. 1 for small input amplitudes V_A.

Fig. 4. Comparison of simulated and calculated results for R_{ds} over input amplitude.

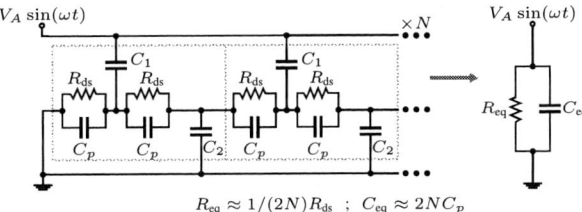

Fig. 5. Simplified circuit model for determining an input impedance approximation.

the output voltage with sub-zero–V_{th} FETs in the linear region to be formulated:

$$V_{out} = 2N \cdot (V_A - V_{drop})$$
$$= -2N \cdot \left(V_{th} + \frac{1}{2}\sqrt{4V_{th}^2 - 2V_A^2}\right) \qquad (6)$$

Again, this approximation is valid only for small input amplitudes of $|V_A| < |V_{th}|$. For larger input signals, the transistor temporarily leaves the linear region, which necessitates another equation. In contrast to the results in [6], the output voltage V_{out} remains independent of the dimensions of the transistor, but is now dependent on its threshold voltage. Although the bulk effect is not considered, the results from the analytical equations are in good agreement with the simulated values, as can be seen in Fig. 3. The simulations were run for an input frequency of 868 MHz to meet the license-free European SRD band specifications. In order to verify the presented equations, all simulations were also carried out for 915 MHz and 1 GHz, which delivered similar results. For simplicity, only the results for 868 MHz are shown.

Amplitudes of up to 40 mV are covered, as for larger values the transistor slowly leaves the linear region. Simulations showed that the behavior reaches a nearly linear trend. However, this effect has not been analyzed further, as this work aimed for maximum sensitivity and low receive-power levels, and therefore only cases in which $|V_A| < |V_{th}|$ were considered. In order to optimize charge pumps for sensitivity, we solve (6) for V_A to make clear the impact of the number of stages N on sensitivity:

$$V_A = \sqrt{-2\frac{V_{out}}{N}V_{th} - \frac{V_{out}^2}{2N^2}} \qquad (7)$$

It can be seen that the input amplitude required[1] for a given output voltage is reduced by $\approx 1/\sqrt{N}$. This means that for each additional stage the improvement in sensitivity decreases. Designs with of more than 6 stages therefore tend to cause more parasitic effects and have reduced sensitivity, negating almost all improvements. This equation is used later

[1]The corresponding receive-power level varies with the input impedance of the charge pump. This is described in detail in Sec. II-D.

for determining the overall sensitivity (including a matching network).

B. Input Impedance

As mentioned in the Tntroduction, the input impedance is dependent on the input amplitude of the charge pump and its current charging state. In order to achieve maximum sensitivity, the matching network should therefore be designed for the case in which proper transmission is achieved with the lowest output voltage. By means of (7), the input amplitude corresponding to the a given output voltage is defined. With the amplitude thus known, the input impedance for designing the matching network can be determined.

To this end, the circuit is transformed to a simplified parallel circuit, as visualized in Fig. 5. First, all transistors are replaced with their effective resistances R_{ds} and their parallel parasitic capacitances C_p. These values are summed to obtain an equivalent resistor R_{eq} and an equivalent capacitor C_{eq}. Examples implementations have shown, the capacitances C_1 and C_2 can be assumed to be ten times larger than the parasitics C_p, which can therefore be ignored when determining the equivalent C_{eq} for the desired frequencies at around 868 MHz. Further $C_{eq} \approx 2NC_p$. Before R_{eq} can be calculated, the small-signal resistances R_{ds} must be determined. R_{ds} can be calculated by

$$R_{ds} = \frac{1}{g_m} = \frac{1}{-K(V_{th} + V_{ds})}, \qquad (8)$$

where V_{ds} has already been defined in (4) and is time-dependent due to the input signal applied and the current charging state. Assuming that the circuit is in the steady state and replacing the sinusoidal component with an equivalent voltage, yields that $V_{ds} = \alpha \cdot V_A - V_{drop}$. In this case, $\alpha \cdot V_A$ is used as the equivalent voltage. The complex calculation of the mean resistance value over one period of the sine-wave signal can be avoided. Using the new expression for V_{ds} gives:

$$R_{DS} \approx \frac{1}{-K\left(V_{th}+\left(\alpha \cdot V_A - \left(\frac{1}{2}\sqrt{4V_{th}^2 - 2V_A^2}+V_{th}+V_A\right)\right)\right)}. \tag{9}$$

The R_{DS} values calculated by (9) are now compared to the simulation results in Fig. 4. The R_{ds} values in the simulation were averaged over time. Setting α to 1—which means that the modulation of the resistance caused by the sine wave is linear—results in a curve that is far from the simulation results. For the curve to fit to the simulated curve, α has to be 0.8, which means that the effective voltage for the sine wave over the resistor is less than its mean value. Note that the simulation showed that resistance and transconductance over a diode-connected FET does not scale linearly with its dimensions for small lengths and widths. In further comparisons, we compensated for this effect by replacing the actual transistor width with an effective width to keep this scaling linear. It should be kept in mind that the actual width drifts by up to 8% for the smallest dimensions possible. The input admittance is then described by:

$$Y_{in} = \frac{1}{R_{eq}} + j\omega C_{eq} \tag{10}$$

with $R_{eq} \approx 1/N \cdot R_{ds}$ and $C_{eq} = 2NC_p$. This admittance is used to calculate the matching network needed for a given antenna. In our case, a typical $50\,\Omega$ resistance is assumed, but first, the resistance R_{ds} is used to determine the settling behaviour of the charge pump.

C. Charge Time and Data Rate

In addition to the formation of the signal envelope from very low input receive-power levels, the dynamic behavior (i.e. the charging and discharging) is important when implementing data rates in the kHz range. Therefore, the charge-time to 90 % is approximated and used as a limit for the maximum achievable data-rate DR $= 1/(2 \cdot T_{charge})$. The discharging behavior is inherently given by the always-on behavior of the sub-zero-V_{th} transistors. Since no transistors that reduce the reverse leakage of the pump are used (as shown in [4]), discharging is actually faster than charging of the circuit and need not be considered at this point.

As shown in [8] and [9], the charging curve of the Dickson charge pump can be modeled by an equivalent RC circuit, as visualized in Fig. 6. In both [8] and [9], an ideal charge-pump behavior was assumed with enough time in each cycle to settle all capacitors; their calculations are therefore not directly applicable to our sub-zero-V_{th} transistors with a sinusoidal RF input. However, we can use their finding that the equivalent resistance for charging R_t is linked to the amount of charge transferred per clock cycle and the number of stages N according to $R_T = N \cdot T/C$. Using this, we can say that an equivalent small-signal resistance R_{ds} across the transistor

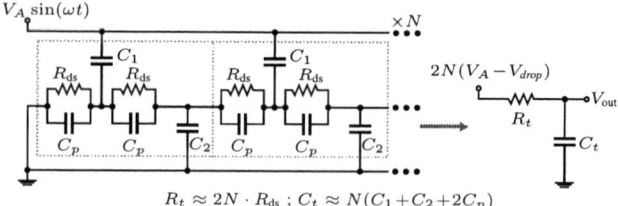

$$R_t \approx 2N \cdot R_{ds}\,;\, C_t \approx N(C_1 + C_2 + 2C_p)$$

Fig. 6. Simplified circuit model for determining a timing constant.

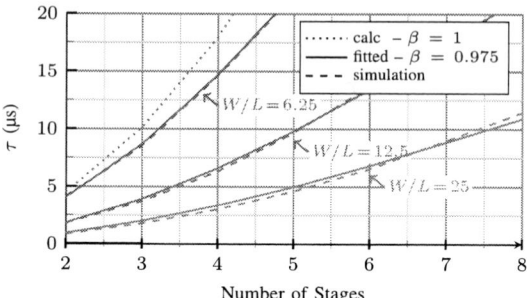

Fig. 7. Timing constant of the described charge pump, over number of stages for various ratios of transistor dimensions and $C_1 = C_2 = 1\,\text{pF}$ and $L = 0.5\,\mu\text{m}$.

(as shown in Fig. 6) limits the amount of current that flows per cycle, and

$$R_t = 2N\frac{T}{C_1} = 2N\frac{V_{ds,m}T}{Q_1} = N\frac{V_{ds,m}}{I_{ds,m}} \approx 2NR_{ds}, \tag{11}$$

where $V_{ds,m}$ and $I_{ds,m}$ are the mean values over one period without any pre-charge of the charge pump. This includes no variations over time and consequently no starting state of the individual stages. If these effects were considered, R_{ds} depend on N. Thus, a constant β is introduced to scale the impact of each stage on the charging time. β is also determined by fitting to the simulation results. The equivalent capacitance C_t (see [8] and [9]) is formed by the total capacitance to be charged. For N stages we can therefore approximate a timing constant by:

$$\tau = R_t C_t = 2N\beta^{2N} \cdot R_{ds} \cdot NC_{stage}. \tag{12}$$

The resistance over a diode-connected transistor is given in (9). Using the input-dependent curve from Fig. 4, one can choose $R_{ds}(V_a = 0)$, as this offers the largest resistance and also describes the situation when there is no pre-charge within the pump. The result should consequently show some minor deviation from the simulation and should lead to a artificially higher timing constant. (12) can now be used to approximate the charge time to 90% with $T_{charge} = 2.3 \cdot \tau$ by including the parasitic capacitance of the stages as $C_{stage} = C_1 + C_2 + 2C_p$. Fig. 7 compares the calculated timing constants with the simulated results for various stages and widths, where capacitors sizes were kept constant. Comparison of the curves shows that the approximation resistance already drifts for more than two stages, but for a β of 0.975 the curves agree well with the simulated results.

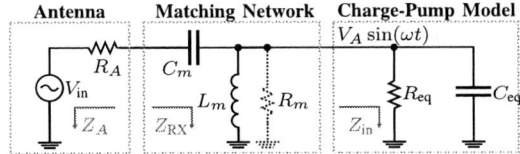

Fig. 8. Model for determining the matching network with passive voltage gain.

D. Matching Network

Completing the front-end requires a matching network to transform a $50\,\Omega$ input impedance of commercially available ceramic chip antennas or similar structures to the input admittance of the receiver circuitry given in (10). Due to the narrow-band system, a simple lossless LC network is used, as shown in Fig. 8. From the requirement that

$$Z_{\text{RX}} = Z_A^* = 50\,\Omega = \frac{1}{j\omega C_m} + j\omega L_m || R_{\text{eq}} || R_m || \frac{1}{j\omega C_{\text{eq}}} \tag{13}$$

two equations for the real- and imaginary part can be derived so that both L_m and C_m can be determined. R_m represents the parasitic parallel resistance of the matching network and results largely from the Q-factor of the inductor ($R_m \approx Q\omega L_m$), where $Q = 40$ for typical off-chip inductors. $\Re(Z_{RX}) = 50\Omega$ is an implicit prerequisite for calculating of L_m and leads to a defined C_m. These two equations are not shown due to their impractical length, but can be derived easily.

Including the bond-wire parasitics slightly improves matching-component estimation. With a typical Q-factor of 40, the 3 dB bandwidth is limited to $\Delta f_{3dB} = 21.7\,\text{MHz}$ at 868 MHz center frequency, which is enough to cover the whole SRD band.

E. Open-Circuit Voltage Sensitivity

Although in this work the term sensitivity is used for simplicity, the technically precise term for describing this kind of receiver circuit is open-circuit voltage sensitivity (OCVS). The OCVS value represents the relationship between input voltage or power level and the achieved output voltage. In this work, the output voltage is assumed to be a defined value (from system simulation) to establish a typical bit-error rate (e.g. less than 10^{-6}), and thus the OCVS for this value is simply referred to as *sensitivity*. The matching network described in Sec. II-D results in a passive voltage boost by the transformation from a small input resistance to a large load resistance. Due to the relation $P = V^2/R$, this passive voltage boost A_{boost} for the amplitudes is described by:

$$A_{\text{boost}} = \frac{V_A}{V_{\text{in}}} = \sqrt{\frac{R_{\text{eq}} || R_m}{R_A}}. \tag{14}$$

This can now be used in combination with (7) to determine the minimum input amplitude for a given output voltage, which is a measure of sensitivity by:

$$V_{\text{in}} = \frac{V_A}{A_{\text{boost}}} = \sqrt{\frac{R_A}{\frac{R_{\text{ds}}}{2N} || R_m}} \cdot \sqrt{-2\frac{V_{\text{out}}}{N} V_{\text{th}} - \frac{V_{\text{out}}^2}{2N^2}} \tag{15}$$

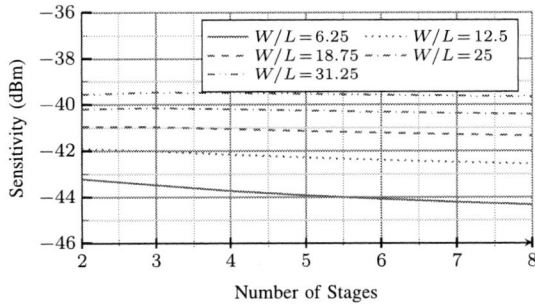

Fig. 9. Sensitivity of charge pump over number of stages for various transistor dimensions (matching network with ideal values and $Q = 40$ included).

For simplicity, R_{ds} and R_m are not replaced with their corresponding equations.

Fig. 9 shows the ideal sensitivity for a 100 mV output calculated using (15) and is referred to a $50\,\Omega$ input impedance. The calculation includes the losses due to an external matching network with a Q-factor of 40. The number of stages and the transistor widths were varied to demonstrate their impacts on sensitivity[2]. The results were verified by simulations. Note that the calculated results include an idealized matching network, which means that no variations or stepping of external components is included. Further no layout effects or wiring parasitics are considered. Thus, the results are only a starting point for designing a receiver. According to Fig. 9, the best sensitivity is achieved by small transistors and a high number of stages, whereas the opposite applies to the charging-time as shown by (12). This leads to a trade-off between data rate and sensitivity. A design flow to find the optimal balance for this trade-off is presented in Sec. III.

F. Noise

Sec. III presents a design strategy that determines the sensitivity of the receiver for a defined output voltage to ensure a particular signal-to-noise ratio (SNR) and a corresponding bit-error rate. For determining the output voltage (e.g. by system simulation), the noise contribution of the envelope detector must be known , which is detailed in the following section. In Sec. II-C the output voltage was described utilizing a simple RC network. A similar approach can be used for describing the noise behavior [10] based on the model shown in Fig. 10. The transistors were replaced with corresponding noise voltage sources and noiseless resistors. The circuit model is now similar to consecutive RC low-pass stages. Due to the frequency behavior of the circuit, the effective noise power can be described by:

$$P_{\text{n,out}} = \int_0^\infty |G(j\omega)|^2 \cdot V_{\text{n,R}} \cdot d\omega, \tag{16}$$

where $V_{n,R}$ stands for the overall thermal noise contribution caused by the transistors and shaped by the transfer function

[2]The data rate is not fixed in the graph; it is assumed that the data rate is set to a maximum so that 90 mV is reached within the charging cycle

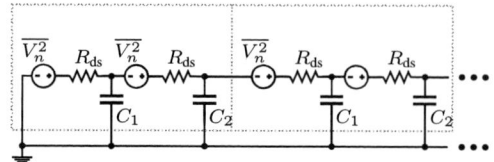

Fig. 10. Model for determining the output noise of the charge pump.

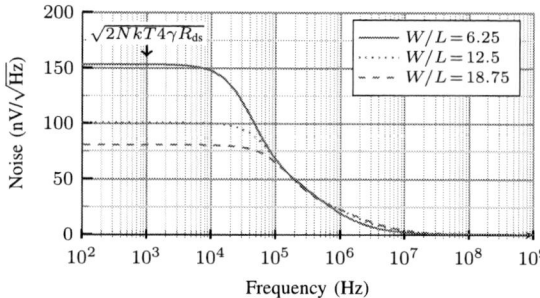

Fig. 11. Noise behavior of the simulated charge pump vs. frequency.

$G(j\omega)$ of the consecutive RC stages. To gain additional insights, the frequency-depended noise contribution at the output of the simulated charge pump for various W/L ratios is shown in Fig. 11. For simplicity C_2 was set euqal to C_1. The overall noise amplitude $V_{n,R}$ is described by summing up the equivalent noise-voltage sources of the transistors with $2NkT4\gamma R_{ds}$. This is also indicated in Fig. 11. Here γ is a parameter that depends on the technology, the transistor sizes and the operating region. In our case, it was determined to be $2/3$. Simulation and numerical integration showed that the integral over G is approximately $(4R_x C_1)^{-1}$ with $R_x = N \cdot R_{ds}$. For the output noise follows that:

$$P_{n,out} = \frac{1}{8NR_{ds}C_1} \cdot 4kT\gamma 2NR_{ds} = \frac{\gamma kT}{C_1}. \quad (17)$$

This expression can now be used to determine an effective noise amplitude for further sensitivity or system-design considerations. Note that the 1/f noise caused by the transistors is not taken into account.

G. Limitations of Modeling

The circuit modeling results (see Sec. II) are in fairly good agreement with the simulation results. However a few aspects that would lead to more accurate results, such as the bulk effect, are not considered. For the simulation shown, a mean threshold voltage was determined. In fact this voltage also depends on the bulk-to-source potential and varies with time and number of stages (as subsequent stages offer a higher source-to-bulk voltage). This affects the output voltage directly via (6) and indirectly by changing the input resistance via (9).

Next is considering the limitations in the linear operating region. As the input voltage of the charge pump exceeds the threshold voltage, the transistor temporarily leaves the linear

region and (6) is no longer valid. Therefore, (6) can only be used for minimum input levels.

Further aspects that have already been mentioned are the layout parasitics and process variations. The effective input capacitance is determined primarily by the low gate capacitance of the transistor and secondarily by the wiring parasitics. As these are hard to cover, one would still need an extensive parasitic extraction to determine the correct impedance and yet matching network. Process variations aggravate this problem, as small transistor sizes cause large variations in threshold voltage and gate capacitance.

Potential improvement due to tapering of the capacitances C_1 and C_2 throughout the stages is also not considered. While noise behavior and impedance values can easily be adapted by using the models shown in Fig. 5 and 10, determining the effect on the charge time is not trivial. As tapering schemes affect overall sensitivity and charge time differently. Determining the ideal tapering scheme for each optimization goal would require a extensive analysis. Since this work aimed to provide a flexible and easy-to-implement design flow, the standard case without tapering as a trade-off between performance and implementability was chosen.

The matching networks used for the calculation are also assumed to be ideal, which means that precise values are used without considering common E-series and tolerances. For some cases in which the resistive part of Z_{in} is high, external capacitances of below 100 fF with an accuracy below 5% would be needed, which is difficult to realize and impractical for high-quantity application.

Despite these limitations, the formulas presented remain a good starting point for a layout design process, that leads to a final optimization that covers all parasitics. They allow the first evaluation and simulation step to be skipped, which saves valuable time when realizin,g for example, a receiver for ASK demodulation. The design strategy presented in the following section further speeds up the design flow.

III. DESIGN STRATEGY AND PROPOSED RECEIVER

This section presents an exemplary design flow for realizing an OOK based wake-up receiver. First, the example receiver architecture is explained, and then the design flow is described by showing how the parameters for the receiver front-end were derived using the equations introduced in the previous section.

Fig. 12 shows a schematic of the example receiver. The front-end consists of an antenna, a matching network and an envelope detector, as described in Sec. II. The output is then filtered by an RC low-pass in order to determine a moving average V_{avg} of the received signal. In the example, a sampling rate of 32 kHz was aimed for, which requires a large chip area to realize a low corner frequency of the low-pass, while offering a passive design without any further adjustments. The generated average voltage V_{avg} is then compared to the envelope signal V_{out} by a comparator, to generate a digitized value. The comparator (see Fig. 13) is realized as a dynamic latched variant because this type has the lowest power consumption [11]. However, this also results in the highest kickback noise,

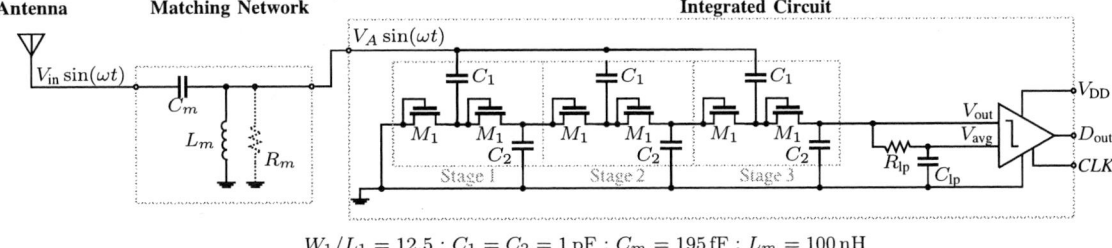

$$W_1/L_1 = 12.5 \ ; \ C_1 = C_2 = 1\,\text{pF} \ ; \ C_m = 195\,\text{fF} \ ; \ L_m = 100\,\text{nH}$$

Fig. 12. Schematic of a simple receiver realized by means of the proposed flow.

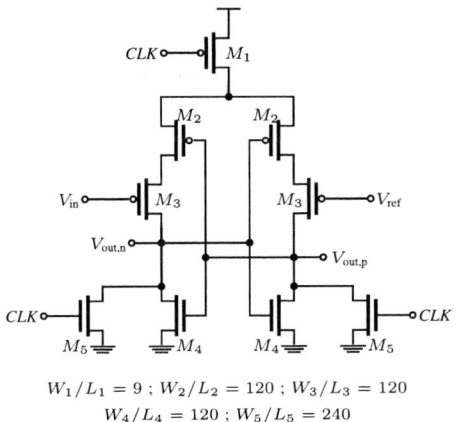

$$W_1/L_1 = 9 \ ; \ W_2/L_2 = 120 \ ; \ W_3/L_3 = 120$$
$$W_4/L_4 = 120 \ ; \ W_5/L_5 = 240$$

Fig. 13. Schematic of the dynamic latched comparator used for the example receiver.

which should be taken into account when determining the parameters for the charge pump. The comparator layout was designed and a Monte Carlo analysis that considered layout parasitics, process and device variations, conducted, which gave an offset voltage of $\sigma \approx 5.5\,\text{mV}$ and a parasitic input capacitance of $C_{\text{comp}} \approx 20\,\text{fF}$. After the digitalization by the comparator, further signal processing, such as averaging or correlation to a known pattern, can be implemented, but this concerns mainly the overall system design and happens in the digital domain. Our approach simplifies the problem for the analog part by assuming a coherent reception of an OOK modulated 868 MHz signal at a fixed sampling rate of 32 kHz. In this setup, the only power-consuming part is the clocked comparator and the downstream digital-signal-processing. The single-bit comparator enables detection of amplitude-modulated signals. In this case, an OOK modulated signal with a specified mean value is assumed.

Having defined the architecture we can now demonstrate the design flow.

Step 1: Determine System Parameters
To start, a system simulation was performed to determine a sampling rate and required SNR that correspond to the modulation scheme and signal processing to be implemented. For the proposed receiver a sampling rate of 32 kHz was set. As described above–and according to theory [12]–a sufficiently low

bit error rate for coherent OOK-modulated signals is achieved with an SNR greater than 20 dB. With this information the modulation depth for the comparator input (referred to as V_{out} in Fig. 12) can be determined to establish the desired minimum SNR. The minimum modulation depth—which in this case is equal to the minimum desired output voltage of the charge pump V_{out}—should be chosen such that this is large enough to cover the uncertainty of average value V_{avg} and the offset of the comparator within the margin between average value and modulation depth. Further, the charge-pump noise and kickback noise should be covered. In the example receiver we chose 100 mV as a starting point. This resulted in a 50 mV margin from the mean value for each modulation point. This margin is then reduced by $3\sigma \approx 16.5\,\text{mV}$ offset voltage and 5 mV uncertainty, which leaves a margin of 28.5 mV. Now capacitor C_1 must be chosen according to the kickback and thermal noise contribution. We selected 1 pF for capacitor C_1, which led to to a comparator kickback noise of $V_{\text{kick}} = V_{\text{DD}}\frac{C_{\text{comp}}}{C_1} = 24\,\text{mV}$. and–according to (17)–a noise amplitude of $53\,\mu\text{V}$. Comparing the remaining 4.5 mV margin after subtracting the kickback noise with the noise amplitude led to a signal-to-noise ratio of more than 20 dB and a sufficiently low bit-error rate [12]. This short calculation gives an estimate and leaves some additional margin for error. For a more detailed result at this point, a system simulation is recommended.

Step 2: Determine Physical Parameters.
To achieve more than 98% of the output signal in half a clock cycle at 32 kHz, we set τ to be less than $5\,\mu\text{s}$. As the values for the capacitances were chosen based on the considerations detailed above, the results of modeling the charge pump system presented in Figs. 7 and 9 can be used to determine the number of stages required and the corresponding width of the transistors. As shown in (15), the smallest width possible would result in the best sensitivity, but as mentioned in Sec. II-G this would result in a tough to realize matching network. In our case, we suggest choosing a transistor ratio of $W/L = 12.5$ and three stages, resulting in a calculated sensitivity of –43 dBm and leads to $L_m = 124\,\text{nH}$ and $C_m = 190\,\text{fF}$.

Step 3: Layout and Revise Parameters.
The layout based on the values determined in our example is shown in Fig. 14. Additional filling structures, over-voltage

protection and bonding pads with included ESD protection were considered during analysis, but were removed from the figure to maintain simplicity.Based on parametrized blocks, creation of layout configurations is fast and can even be automated. The design occupies a total area of 0.03 mm^2. The finished layout should then be used to extract all parasitic effects, including bond pads and -wires. This leads to a revised value for the input impedance, and with the equations from Sec. II-D the matching components should be recalculated. In our case C_m is changed to 195 fF and $L_m = 100$ nH. These results are still manageable. The additional layout parasitics and changed matching network caused the final sensitivity to decrease to −40.5 dBm, and a τ of 4.8 µs was simulated. In this case, the results were still feasible; if the matching network becomes to difficult to realize, or τ does not meet the predefined requirements, the process can be restarted at Step 2 and the choices for W and N revised or the layout parasitics reduced if necessary.

IV. Conclusion

This paper has presents various models for describing a Dickson charge-pump-based receiver design with sub-zero-V_{th} transistors, for achieving a rapid design-flow. All models and approximations were verified by extensive simulations, and result is a fair approximation of the receiver. We have described the validity of the formulas presented and shown a design strategy for an example receiver.

Further improvements could be achieved by, for instance, setting up on the findings from Sec. II-C. Reducing capacitor size for each consecutive stage would allow a shorter settling time to be realized (as shown in [13]). However, this also affects noise behavior and input impedance. Equation (12) would have to be adapted to this scenario or could be used to determine a starting point for a design, that is improved with the technique described in [13].

We have presented and example receiver design to illustrate the design flow and the ease with which all required variables, such as transistor width, number of stages and capacitor sizes, can be determined. As a result, a ready-to-use layout is obtained. The design strategy presented in Sec. III also enables the realization of an automated flow. Layouts for pre-parametrized cells for capacitors and transistors could be used to generate a design starting point, with the help of the equations presented in Sec. II. The resulting first layout design can be used to extract and simulate all parasitics and refine the corresponding design values for a more accurate description of the final receiver design.

With the proposed flow, a ready-to-use layout for a receiver is obtained that occupies a total area of 0.03 mm^2 and provides a simulated sensitivity of −40.5 dBm in a 1P6M 180 nm technology.

Acknowledgment

We thank Ansys for providing their VeloceRF and RaptorX tools, which were used for parasitic simulation of the elaborated designs.

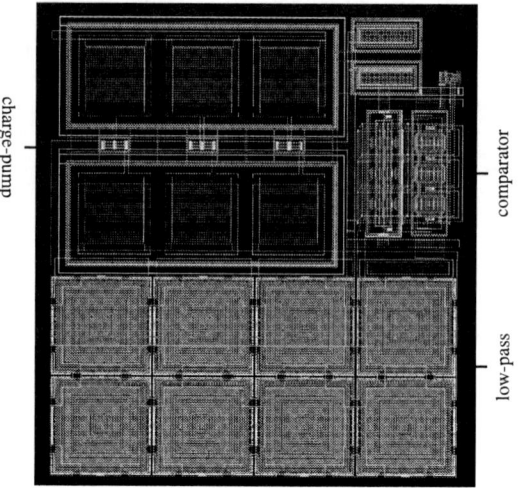

Fig. 14. Layout scheme for the charge pump of the 160×180 µm example receiver in Sec. III.

References

[1] A. Ballo, A. D. Grasso, and G. Palumbo, "A review of charge pump topologies for the power management of IoT nodes," *MDPI Electronics*, vol. 8, no. 5, p. 480, 2019.

[2] S. M. Noghabaei, R. L. Radin, Y. Savaria, and M. Sawan, "A high-efficiency ultra-low-power CMOS rectifier for RF energy harvesting applications," in *Proc. IEEE Int. Symp. Circuits and Systems (ISCAS)*, May 2018.

[3] J. Dickson, "On-chip high-voltage generation in MNOS integrated circuits using an improved voltage multiplier technique," *IEEE Journal of Solid-State Circuits*, vol. 11, pp. 374–378, 1976.

[4] S. Schmickl, T. Faseth, and H. Pretl, "An RF-energy harvester and IR-UWB transmitter for ultra-low-power battery-less biosensors," *IEEE Transactions on Circuits and Systems I: Regular Papers*, vol. PP, 2020.

[5] J. Moody and S. M. Bowers, "Triode-mode envelope detectors for near zero power wake-up receivers," in *Proc. 2019 IEEE MTT-S International Microwave Symposium (IMS)*. IEEE, 2019, pp. 1499–1502.

[6] S. Oh and D. D. Wentzloff, "A 32dBm sensitivity RF power harvester in 130nm CMOS," *2012 IEEE Radio Frequency Integrated Circuits Symposium*, pp. 483–486, 2012.

[7] J. Yi, W.-H. Ki, and C.-Y. Tsui, "Analysis and design strategy of UHF micro-power CMOS rectifiers for micro-sensor and RFID applications," *IEEE Transactions on Circuits and Systems I: Regular Papers*, vol. 54, no. 1, pp. 153–166, jan 2007.

[8] A. Ballo, A. D. Grasso, G. Giustolisi, and G. Palumbo, "Optimized charge pump with clock booster for reduced rise time or silicon area," *IEEE Transactions on Circuits and Systems II: Express Briefs*, vol. 66, pp. 1977–1981, 2019.

[9] G. Palumbo, N. Barniol, and M. Bethaoui, "Improved behavioral and design model of an nth-order charge pump," *IEEE Transactions on Circuits and Systems I: Fundamental Theory and Applications*, vol. 47, pp. 264–268, 2000.

[10] P. Bassirian, J. Moody, and S. M. Bowers, "Analysis of quadratic dickson based envelope detectors for ioe sensor node applications," in *Proc. IEEE MTT-S Int. Microwave Symp. (IMS)*, 2017, pp. 215–218.

[11] P. Figueiredo and J. Vital, "Kickback noise reduction techniques for cmos latched comparators," *IEEE Transactions on Circuits and Systems II: Express Briefs*, vol. 53, pp. 541–545, 2006.

[12] A. F. Molisch, *Wireless communications*. John Wiley & Sons, 2012.

[13] A. Saeed, S. Ibrahim, and H. F. Ragai, "A sizing methodology for rise-time minimization of dickson charge pumps with capacitive loads," *IEEE Transactions on Circuits and Systems II: Express Briefs*, pp. 1202–1206, 2017.

Sorting Network based Full Adders for QDI Circuits

Florian Huemer and Andreas Steininger

Institute for Computer Engineering, TU Wien, Vienna, Austria

{fhuemer,steininger}@ecs.tuwien.ac.at

Abstract—**In this paper we focus on the efficient design of ripple-carry adders (RCAs) in quasi delay-insensitive (QDI) logic. We review approaches from the literature that are suitable for implementation in modern CMOS technology and identify their weaknesses. Based on these insights we propose novel full adder designs that are built around binary sorting networks. In order to maintain the timing robustness of QDI logic, a thorough management of gate orphans is necessary. Our analysis gives evidence that the proposed circuits are more area efficient than existing approaches, while still maintaining a comparable performance in terms of average delay.**

Index Terms—**asynchronous circuits, QDI, full adders, sorting networks**

I. INTRODUCTION

Adders are very versatile generic functions that are not only heavily used in arithmetic addition and subtraction operations, but also for comparisons, multiply-add operations, counters, and many more. Even though many sophisticated architectural enhancements exist, the ripple-carry adder is still a popular function, and even more so its constituent full adder (FA) primitive which will, therefore, be in our focus.

Addition is an often cited example for an operation with input data dependent duration. Therefore its synchronous implementation based on worst case timing assumptions results in particularly high performance penalties, as in many cases the computation is finished way earlier than in the worst case. With a fixed clock, however, there is no way to leverage that early completion.

Asynchronous logic, in contrast, promises to offer the timing flexibility required for that. Specifically the QDI design style is able to adapt its speed of operation to the current situation (actual circuit delays determined by process variations, supply voltage, temperature, computational path determined by input data), rather than being bound to a pessimistic worst case. As will be shown in the paper, among the options for implementing FAs in QDI, some are more suitable for adapting to the data dependent delays, while others are not. We will analyze existing approaches with respect to their average speed (i.e., for random uniformly distributed, not worst-case input data) and their costs in terms of area, and we will propose and assess new, enhanced implementations. However, in this paper we restrict our investigation to logic styles that only use basic gates readily available in standard libraries (with no more than 3 inputs). The only exception is the 2-input C gate [1]

This research was partially supported by the project ENROL (grant I 3485-N31) of the Austrian Science Fund (FWF).

as it is indispensable in most asynchronous and QDI design styles[1]. Hence, this restriction excludes design styles based on precharged/domino logic [2] as well as Null Convention Logic (NCL), which relies on threshold gates [3].

In the next section we will give a brief review of existing solutions from the literature. Then, in Sec. III we will present background on the basic concepts of asynchronous design that will be needed further on, followed by an overview of implementation templates for QDI combinational blocks in Sec. IV. Sec. V will then present our own new designs. In the elaboration of the three different variants we propose, important aspects are the degree of concurrency in combination with the decision between strongly and weakly indicating behavior, the management of orphans, and the identification and optimization of the paths that are critical to performance. All these need to be carefully balanced, while a rigorous elimination of potential orphans must be ensured. To explore the efficiency of our solutions, we compare their speed and area with the approaches from literature and discuss the differences in Sec. VI, before concluding the paper in Sec. VII.

The aim of this paper is not to give a definite answer to the question which adder design is the best. It aims to give an overview of the available designs and present some estimates on their area overhead and performance. The selection of a particular design depends on the available target technology, hence we keep our analysis on a more abstract level.

II. RELATED WORK

Since adders are very fundamental building blocks, there have been many approaches to build them in a delay-insensitive fashion. Many of them focus on the architectural optimization, like using carry-lookahead or similar strategies [4], [5] , while only few perform a comparison of implementation variants on the level of asynchronous circuit implementation. One example for the latter is [6] where adders using various choices for protocol (return-to-zero vs. return-to-one), indication behavior (strongly vs. weakly) and the use of timing assumptions are compared side by side.

In our paper the scope is restricted to the FA block itself and its use in RCAs. However, the fundamental nature of the FA and its use in many advanced circuits, aids in giving our comparison results a higher level of generality . Unlike [6] we stricly focus on QDI implementations and restrict the discussion to the return-to-zero protocol (extension to return-to-one are straight-forward). We also remain with our

---

[1] We assume that a single custom cell for the 2-input C gate is provided as a minimum investment for facilitating a QDI implementation

978-1-7281-8494-4/20 $31.00 © 2020 IEEE

implementations with static gates and do not include dynamic circuit techniques, like [7] and [8]; neither do we consider threshold gate-based logic [3].

The key aspects for the implementations presented in this paper are degree of parallelization, sharing of sub-functions, control of orphan transitions, and degree of use of non-standard library elements like Muller C-gate.

III. BACKGROUND ON ASYNCHRONOUS DESIGN

Where synchronous circuits use the clock signal to control data transfer between storage elements (e.g., pipeline stages), asynchronous circuits use some form of closed-loop handshaking protocol. This handshake mechanism (usually) involves two signals, called request (req) and acknowledgment (ack). The data source uses the req signal to indicate the availability of new data to the sink, which then acknowledges the reception using the ack signal[2]. Depending on the number of transitions on these two wires for one complete handshaking cycle, the protocol can be classified as 2-phase or 4-phase. A 4-phase handshake is defined by the switching sequence $req\uparrow$, $ack\uparrow$, $req\downarrow$, $ack\downarrow$, where the arrow symbols denote rising and falling transitions, respectively. Hence, the handshake signals always start and end the handshaking cycle with the same logic value. In contrast to that, 2-phase protocols only use two transitions and leave the handshake signals in the opposite logic state. This means that e.g., $req\uparrow$, $ack\uparrow$ or $req\downarrow$, $ack\downarrow$ both constitute complete 2-phase handshakes.

The described request mechanism need not to be implemented as a dedicated req wire. It is also possible to implicitly encode the request into the transmitted data using some delay-insensitive (DI) code [9], which leads to the class of QDI circuits [10]. The sink then has to use a completion detector (CD) to decide when the received data is complete (i.e., valid) and can thus be consumed and acknowledged. QDI circuits can also be implemented using 2-phase or 4-phase protocols. However, for circuits that actually process data (in contrast to just transmitting, e.g., [11]), practically only the 4-phase variant is relevant. Hence, in this paper we focus on this style. For the same reason we will only consider the dual-rail (DR) data encoding as DI code, even though a variety of DI codes are applicable [12]. The DR code encodes each bit using two rails (i.e., wires) which we refer to as the true and false rail. For a DR bit d, we denote these rails by $d.t$ for the true rail and $d.f$ for the false rail. In every handshaking cycle all data rails start out in a low state (spacer) and only one rail of each DR bit can go high. After acknowledgment all rails switch back to low and the whole process start over. Hence the receiver can simply use an OR gate to check for completion on each received DR bit. The individual phases of this protocol are referred to as *null* (or *spacer*) and *data* phase.

Fig. 1 shows an example timing diagram of the transmission of two DR bits d_0 and d_1. The order of the transmitted bits in the figure is given by $(d_1.t, d_1.f, d_0.t, d_0.f)$. The transmitted

---

[2]This explanation assumes push channels, pull channels will not be considered in this work.

binary data (d_1, d_0) can hence be decoded to $(0, 0)$ followed by $(1, 0)$.

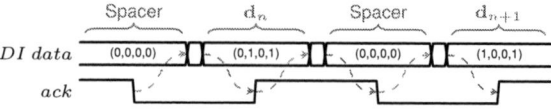

Fig. 1. 4-phase Protocol

It is important to stress that the order in which the transitions on the individual data rails arrive at the receiver does not matter. This is an important property of a DI code. Furthermore, the individual gate and wire delays in a QDI circuit can vary arbitrarily and the circuit is still guaranteed to work correctly. This, in turn, is an important property of a DI circuit. The only restriction on the delays is the so called isochronic fork constraint, which demands that for certain wire forks the delays of the individual paths behind the fork must be equal. This is also the reason why this class of circuits is referred to as "only" *quasi* DI.

Fig. 2 shows a Weak-Conditioned Half Buffer (WCHB) based QDI pipeline containing some combinational logic [2]. The storage elements in the buffers are C gates. The Muller C-element, or short C gate, is a fundamental gate in asynchronous logic. Its function is to output the logic level seen at its inputs when these match, and to retain the last valid output state otherwise. It can hence also be viewed as an AND gate with hysteresis. Its capability of retaining the last valid output state implies some form of storage; consequently the C gate has substantially higher area cost than a conventional AND.

Each input rail is connected to one such C gate. The other C gate input is controlled by the CD of the succeeding stage and, depending on the implementation of the combinational logic, an internal completion signal of the function block ($done$). This structure ensures that alternating waves of data and spacers can traverse the pipeline. More specifically, the CD indicates to the predecessor stage that the data (null value) provided to the input of the stage has been properly processed and stored, and can hence be replaced by the next item.

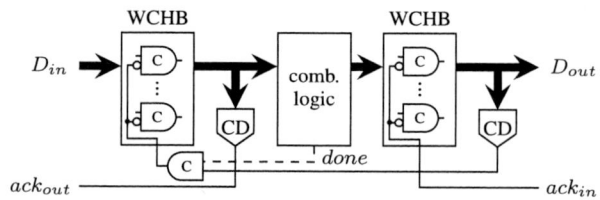

Fig. 2. QDI pipeline structure with combinational logic

The combinational logic operates on DR signals and must hence be implemented in a logic style that ensures the preservation of the circuit's QDI properties. In particular this means that the last (DR) output of the combinational logic may only transition to the data phase after *all* inputs are in the data phase. Conversely, the last output must only transition to the spacer phase if all inputs have switched to the spacer. For this protocol to work properly, each rail is allowed to make only one transition (if any) per phase, which forbids the occurrence of glitches anywhere in the circuit. Furthermore, it must be guaranteed that the circuit is free from gate orphans [3]. A gate orphan is defined as a gate in a circuit that would change its output value for some particular input pattern, but whose output does not have any impact on the primary (observable) outputs of the circuit. Since in the QDI timing model gate delays may be arbitrarily long, such orphan transitions can potentially interfere with following data phases and cause a malfunction in the circuit. Hence for QDI circuits it must be possible to determine if a (combinational) circuit has finished its computation by just observing its outputs.

If this requirement cannot be satisfied for a combinational circuit by observing its data outputs alone, an additional control signal (*done*) may be introduced which "collects" all these orphan transitions from the "hidden" internal nodes, essentially resembling an internal completion signal. This *done* signal is combined with the acknowledgment signal, such that the circuit can only proceed when the combinational logic has stabilized, i.e., all gates that needed to transition for the particular input data (or the spacer), actually finished transitioning (see Fig. 2).

IV. QDI FUNCTION BLOCK DESIGN

One possibility to do implement QDI function blocks is the Delay-Insensitive Minterm Synthesis (DIMS) design style [13]. DIMS uses an array of C gates to exclusively map every possible (valid) DR input data word to a dedicated signal (one-hot code). In a second stage OR gates map this code to the desired output signals. The actual logical function depends only on these OR gates. Fig. 3a demonstrates this approach for a DR 2-input AND gate. Notice that circuits produced with the DIMS design style don't contain orphans and thus there is no need for an internal CD.

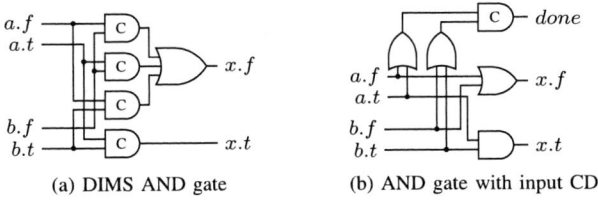

(a) DIMS AND gate   (b) AND gate with input CD

Fig. 3. QDI DR AND gates

It is apparent that this design style does not scale well with the number of inputs and quickly leads to rather large circuits, hence many different styles and optimizations have been proposed. A different implementation template for combinational circuits that does not rely on C gates to implement the actual boolean function was proposed in [14] and further improved in [15], where it is called NCL-X. However, we will present it here in a slightly different way in order to better fit our explanations and narrative. Fig. 3b shows an AND gate implemented in this style. As can be seen the output signals $x.t$ and $x.f$ of the gate are generated using an AND and an

OR gate. Without further precautions a DR AND gate only comprised of these two gates would violate the QDI properties presented in Sec. III. In a data phase, where one DR input is false (i.e., its false rail is asserted), the circuit immediately produces a valid output signal (with the OR gate asserting $x.f$) without waiting for the second input to become valid. A similar situation arises in the reset phase where the output may transition to the spacer before both inputs did so. For this reason the circuit in Fig. 3b also includes the input CD which generates the *done* signal. Here C gates are still required in order to merge the completion signals for the individual DR bits. However, as shown in [15] in practice this CD can often be shared with the CD of the preceding pipeline buffer.

Finally, let's investigate how a circuit consisting of multiple cascaded gates would be implemented. For that purpose consider the function $y = (a \wedge b) \vee c$. Using the DIMS design style this is quite straight-forward to implement. The DIMS AND gate from Fig. 3a can simply be connected to a DIMS OR gate[3], no special care must be taken to avoid orphans. Note that the given function could also have been implemented using a 3-input DIMS gate. However, such a circuit requires 3-input C gates, which we don't consider available in our target library, and would even have higher area requirements.

Now consider the circuit shown in Fig. 4. Similar to the CD at the inputs of the AND gate in Fig. 3b all three inputs of the this circuit must be equipped with CDs as well. Since the output of the DR AND gate is not used as a primary output of the overall circuit, a CD is required for this intermediate DR signal.

Note that collecting all orphans to produce the *done* signal usually involves a large tree or cascade of C gates (or a multi-input C gate, which is not practically feasible). However, this happens concurrently to the actual computation of the output value, which, compared to the DIMS approach, involves fewer gates, resulting in faster circuits. Furthermore, even with the internal CDs the area overhead can still be significantly lower as we will see in Sec. VI. Also, notice that the *done* signal actually propagates upstream, i.e., its destination is the previous pipeline buffer, where it joins with the acknowledgment signal of the succeeding buffer. However, for this acknowledgment signal to transition the data signals have to traverse the buffer C gates as well as the CD of the associated stage. This is obviously not required for the *done* signal, which further reduces the performance criticality of this signal path.

The FAs that will be presented in the next section rely on the concepts and design styles presented here.

V. QDI FULL ADDERS

A FA, such as the one shown in Fig. 5a, is a digital circuit with three inputs a, b, and c_{in} (carry input) and two outputs s (sum) and c_{out} (carry output). Its purpose is to add the three (single-bit) input signals and produce a 2-bit result consisting of the sum bit (i.e., the LSB of the 2-bit result) and a carry

---

[3]In QDI design signal inversions are achieved by swapping the true and false rail of a DR bit. Hence, by De Morgan's rule a DR OR gate is equivalent to a DR AND gate with all input and output rails swapped.

978-1-7281-8494-4/20 $31.00 © 2020 IEEE

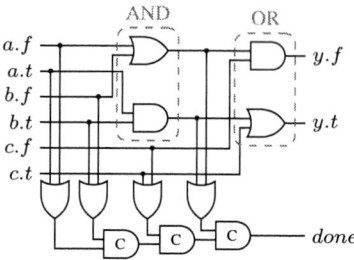

Fig. 4. AND-OR structure with 3 inputs and input/internal CD

output. To add binary numbers of arbitrary width n, a RCA consisting of n FAs (FA$_0$,...,FA$_{n-1}$) can be used. As shown for the example of $n = 2$ in Fig. 5b this is done by connecting the i-th bits of the input vectors $\mathbf{a} = (a_0, ..., a_{n-1})$ and $\mathbf{b} = (b_0, ..., b_{n-1})$ to the a and b inputs of FA$_i$, while the carry output of FA$_i$ is connected to the carry input FA$_{i+1}$ to form a so called carry-chain. The delay of this carry-chain represents the critical path through a RCA in a synchronous circuit. Therefore, its optimization is critical for the attainable speed of the circuit. While the performance of our envisioned asynchronous implementation is not bound to the worst case delay but can benefit from average delays, the optimization of the carry path is still a crucial design goal.

As illustrated in Fig. 5a, a FA can also be viewed as being composed of two half adder (HA) components. A HA just adds two single-bit numbers and is hence considerably simpler. For RCAs that don't require a c_{in} signal the first FA can be replaced by a HA.

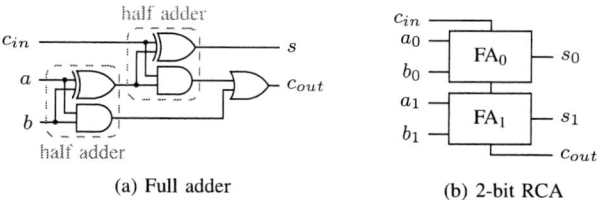

(a) Full adder   (b) 2-bit RCA

Fig. 5. Binary addition circuits

Multi-output QDI function blocks can be classified as strongly and weakly indicating [16]. Since a FA is a multi-output circuit we have to introduce this notion here. A strongly indicating function block only starts switching (any of) its outputs to the data (null) phase when *all* inputs switched to the data (null) phase. Weakly indicating circuits may already change some output data with only a subset of the inputs in a suitable phase. However, in both cases it is guaranteed that the last output switches to the data or null phase only after all inputs switched to the respective phase. Recall from Sec. III that only at this point the CD triggers the change to the next phase at the input.

In the context of QDI FAs in RCAs this is an important performance-determining property. Using strongly indicating FAs the carry signal *always* has to travel through the complete carry-chain, because every FA only starts producing output values when it received its carry input.

To achieve good average performance a weakly indicating FA is preferable, because it shortens the average distance a signal has to travel in the carry-chain: If inputs a and b match (which is the case for statistically 50% of the inputs), then c_{out} is already determined irrespective of the value of c_{in}. This allows an early activation of c_{out} therefore effectively shortening the carry path; in essence it breaks the carry path into segments (depending on the data inputs) that can be processed independently and hence concurrently. This also occurs in synchronous implementations, but there the reliance on the worst-case timing prohibits making use of this effect. QDI circuits, in contrast, can fully benefit from this average case timing – if they allow weak indication.

The first QDI FA we want to discuss is the version using 2-input DIMS gates. A first naive approach to such an implementation is to replace each gate in Fig. 5a with its DIMS version, which each consists of four C gates and one or two OR gates, depending on the boolean function (see Fig. 3a). However, notice that the two pairs of XOR and AND gates (i.e., the two HAs) share the same inputs, meaning that the C gates in their DIMS versions would perform the exact same operation. Hence these gates can be shared between the DIMS AND and XOR gates. A different optimization for this design style called "Input Completeness Relaxation" was proposed by Jeong and Nowick in [17]. This optimization essentially allows to replace the two DIMS AND gates with simpler versions consisting of just an OR and an AND gate, without compromising the QDI properties of the overall circuit. This simpler AND gate is essentially the one shown in Fig. 3b but without the input CD. This circuit is the one we will refer to as DIMS FA. In this paper we don't consider a potential FA variant using a single 3-input DIMS gate. This is because such a solution would require 3-input C gates, which we want to avoid, and would bring little to no benefit when compared to other presented circuits.

Another possibility is to replace every gate in Fig. 5a with its NCL-X equivalent, resulting in the circuit shown in Fig. 6. This circuit needs CDs for the three input signals as well as the three internal signals, which are also depicted in the figure. Notice that the marked gates can be shared between the XOR and AND gates. We will refer to this circuit as NCLX2 FA.

The NCL-X design style can also be applied using 3-input gates, resulting in the circuit shown in Fig. 7, which we will refer to as NCLX3 FA. For every input pattern only one of the AND gates is activated, the OR gates at the output then determine which output rails must be asserted. Since the AND gates switch to zero again as soon as one input is deasserted an input CD is required for this circuit. Using C gates instead of AND gates would yield the 3-input DIMS FA.

In [18] Toms presents a synthesis algorithm for QDI combinational circuits, which yields the FA shown in Fig. 8. For every input value only one of the C gates in the first column and one in the second column switch to one. The circuit does not need an input CD and is strongly indicating.

If this circuit is equipped with an input CD and an internal CD to collect gate orphans after the first column of C gates, the

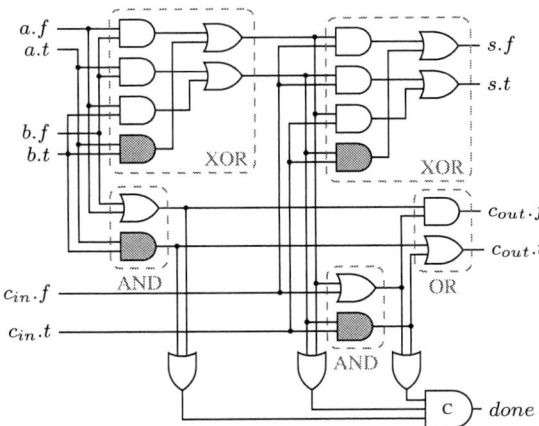

Fig. 6. NCL-X full adder (input CD omitted from figure)

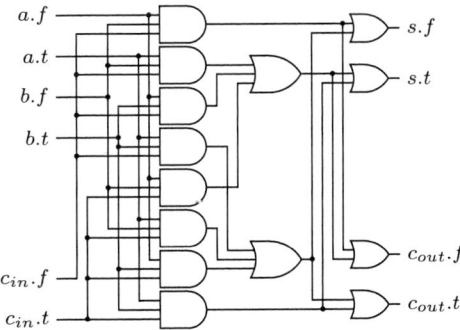

Fig. 7. NCL-X full adder with 3-input gates (input CD omitted from figure)

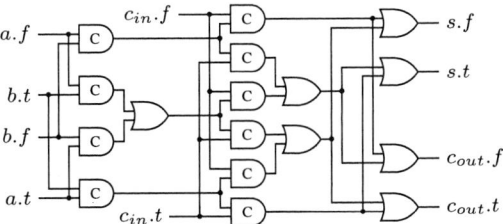

Fig. 8. Toms' full adder [18]

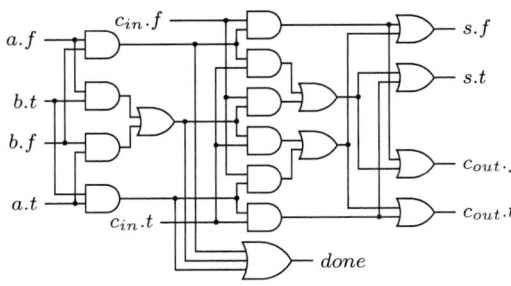

Fig. 9. Modification of Toms' full adder (input CD omitted from figure)

OR gate will mask all transitions on x_0, but at the same time the AND gate will convey them. Conversely, for $x_1 - 0$ the AND gate will mask and the OR gate be permissive. So every input transition does have a visible effect at exactly one of the outputs, and due to the isochronic fork assumption the inputs of the respective masking gate must be stable by then as well, which means that the basic element is orphan-free. The SN, being a network of such elements, is hence orphan-free as well (given that all outputs are observed).

Fig. 11 shows the proposed FA design, which we will refer to as SN FA. The SNs are essentially used to count the number of asserted true and false rails in the set of DR input signals. If any of the SNs detects two asserted inputs (i.e., T_1^3 and T_2^3 switch to one), the respective carry output can be generated, which means that the circuit is weakly indicating. To generate the sum output, all three inputs must be present and one of two conditions must be satisfied. Either one SN detects three asserted inputs, which means that the inputs are either all true (i.e., their sum is one) or all false (i.e., the sum is zero). The other possibility is that one SN detects two asserted inputs and the other SN one. If two true rails and one false rails are detected the sum is zero in the complementary case it is one.

It is important to point out that this circuit does not need input or internal CDs. In the first place, as outlined above, binary SNs don't contain gate orphans. For every combination of input values exactly three outputs of the combined set of outputs of both SNs will transition to one. When all those

C gates can be replaced with AND gates. The resulting circuit, which we will refer to as TomsX FA, is shown in Fig. 9. Note that this circuit can no longer be classified as strictly strongly indicating. While it still only generates output data on s and c_{out} when all three input values have switched to the data phase, the *done* output may be asserted with only two inputs present. Moreover, when one input switches to the null phase the outputs will immediately switch to the null phase as well, only the *done* signal switches to low when all inputs are in the null phase. This behavior also applies to the NCLX3 FA.

Finally we want to present our own designs for FAs based on the use of binary sorting networks (SNs). A binary SN is a circuit, that presented with a binary vector at its input (e.g., 101) produces a sorted binary output vector (e.g., 110). This result can also be viewed as a unary encoded representation of the number of ones in the input vector (i.e., its Hamming weight). In QDI design these circuits can be used to implement efficient CDs for various DI codes [12].

Fig. 10 shows a 3-input/3-output binary SN T^3, which will be used in the following circuits. The output T_i^3 of a T^3 SN switches to one when at least i inputs are asserted.

Note carefully that this circuit does not contain orphans: The basic sorting element comprises an AND and an OR gate, which are both sourced by the same two signals (those to be sorted), like x_0 and x_1 on the left. Hence for $x_1 = 1$ the

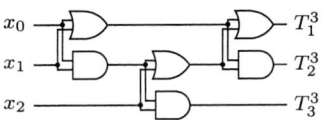

Fig. 10. 3-input binary sorting network

978-1-7281-8494-4/20 $31.00 © 2020 IEEE

three intermediate transitions are involved in the generation of output values it is clear that there are no more orphan transitions that could happen at an output of a T^3. After the last output transitioned to data or null, one can be sure that all switching activity inside the circuit has completed.

Specifically, if on one SN T_1^3 is the only activated output (case I), it takes T_2^3 from the other side to activate the respective sum rail (according to our assumption, the lower input of the OR must be 0). The T_1^3 on the side where T_2^3 is asserted, is required to activate the corresponding carry rail. In case of all 3 outputs activated on one SN (case II), the respective carry rail collects the transitions from T_1^3 and T_2^3, while the transition at T_3^3 is observed at the respective sum rail. Hence, it is clear that in both cases all internal transitions are observed.

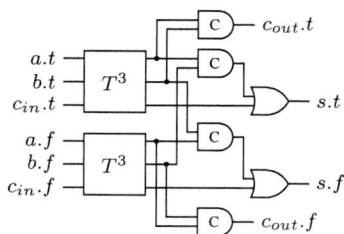

Fig. 11. Sorting network based full adder

Notice that for generating the carry rail $c_{out}.x$ the corresponding output T_2^3 alone would be sufficient, as $T_2^3 = 1$ implies $T_1^3 = 1$. In fact the C gates joining these two signals are only there to keep the T_1^3 signals involved, and specifically avoid introducing gate orphans when switching to the null phase. These additional C gates in the carry path, however, introduce additional delay in the carry chain of a RCA, which is exactly where we need to optimize.

The circuit shown in Fig. 12, which we refer to as the SN Fast Carry (SNFC) FA, improves this issue by removing the C gate from the carry path. However, to still keep the T_1^3 outputs involved in generating an output signal in the case where both SNs assert this output, additional C gates have to be introduced in the signal paths generating the sum output.

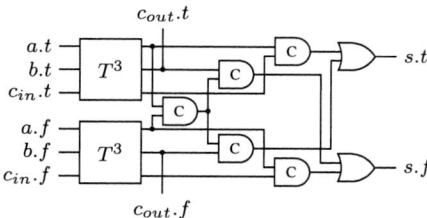

Fig. 12. Sorting network based full adder with fast carry generation

The other possibility to create a fast carry path, but without compromising on the sum output delay, is presented in Fig. 13, which we refer to as the SNX FA. This FA variant does not use C gates to implement the actual boolean function and directly uses the T_2^3 outputs for c_{out}. In order for this to

work correctly, an internal CD must be added to collect all orphan transitions that may arise. Having thus again ensured that all three output transitions of the two T^3 are involved in generating the outputs for every possible input pattern, the circuit still does not require input CDs.

Note carefully, that we have now made the carry generation fast and by removing the C gates from the data path also improved the sum output delay. This optimizations came at the price of introducing the *done* signal, which, for some input patterns, resembles the slowest path in the cell now. Recall, however, that *done* is moving upstream, thus bypassing the next stage and its CD, and hence has some delay margin.

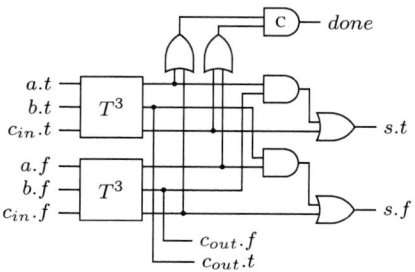

Fig. 13. Sorting network based full adder with internal CD

Table I shows an overview of the properties of the FA circuits presented in this section. Keep in mind that if a FA needs an input CD then every internal carry signal along the carry chain of a RCA also needs a CD.

TABLE I
QDI FULL ADDER PROPERTIES

Circuit	Indication	Input CD	Internal CD
DIMS	weak	no	no
NCLX2	weak	yes	yes
NCLX3	full[4]	yes	no
Toms	full	no	no
TomsX	full[4]	yes	yes
SN	weak	no	no
SNFC	weak	no	no
SNX	weak	no	yes

VI. EVALUATION AND RESULTS

This section evaluates and compares the presented FA circuits in terms of their area overhead and performance.

A. Area Analysis

For the area analysis, we optimized the individual FAs using standard CMOS optimizations, reducing the number of non-inverting gates. As mentioned before we don't target any specific gate library with this analysis. Hence, for the evaluation we assume an area requirement of 1 gate equivalent (GE) for 2-input NAND/NOR gates and 0.5 GEs for inverters. C gates are assumed to require 2GE, which, depending on the actual

---

[4]Weakly indicating on the data rails when switching to the spacer, see explanation above.

implementation may even be considered generous [1]. Hence our analysis only considers the gate costs, other parameters like drive strength/fanout or routing are not considered.

First we consider a classic RCA with a varying data width of 4, 8 and 16 bits and carry input and output signals. Fig. 14a shows the area overheads of the presented implementation variants relative to the DIMS version, which we will use as the baseline for this analysis. The stacked bars in the figure indicate the proportion of the hardware dedicated to C gates (upper part) and combinational gates (lower part). Note that the cost for input CDs (needed for the NCLX2/3 and TomsX versions) is not included in these results, since we assume that the adder is directly fed from a buffer stage (see Fig. 2).[5] It can be seen that the SN based FAs all have a very competitive area footprint. The fact that the NCLX3 FA performs best in this analysis is a little misleading. From the discussed circuits, this variant has the highest fanout on its input signals, which might require additional drivers, and quite complex routing requirements which are not factored in here. Unlike all other FAs presented in this paper the maximum fanout for any signal in the SN based FAs is three.

For the second evaluation we implemented an adder tree consisting of two levels of RCAs adding four numbers. Here we used half-adders for the lower-most bits of the individual RCAs, where a NCLX-style HA was used for the NCLX2/3, TomsX and SNX versions, while the other used DIMS HAs. Fig. 14b shows the results. Since some circuits need CDs at the input of the second-level adder, the NCLX2/3 and TomsX version perform worse than in the previous test. In particular the SNX FA now even surpasses the NCLX3 version. It is also worth noting that with the exception of the NCLX3 FA the SNX version requires the least amount of C gates. This means that if we would assume a more expensive C gate implementation, the SNX version would perform even better.

B. Performance

For the performance evaluation of the presented circuits we derive a simple mathematical model that describes their behavior in a RCA with input data width n. For that purpose we assume that all inputs arrive at the RCA simultaneously and that their values are uniformly distributed (i.e., every input data bit is equally likely to be zero or one regardless of the other input bits).

Let us define the following delay parameters.

- Early carry delay Δ_E: The delay to generate c_{out} out of the inputs a and b for the case where $a = b$ (this only applies to weakly indicating FAs).
- Late carry delay Δ_L: The delay to generate c_{out} when c_{in} arrives, assuming that a and b have already propagated through the circuit as far as possible. Hence internal nodes only dependent on a and b have already settled to a value.

---

[5]Note, that this assumption combined with the "Input Completeness Relaxation" approach from [17] cloud be used to further optimize the DIMS and Toms FA variants resulting in a hybrid NCL-X/DIMS solution, which relies on an input CD provided by the source buffer. However, such specific circuit-level optimizations go beyond the scope of this paper, hence we restrict the discussion to the "pure" DIMS solution.

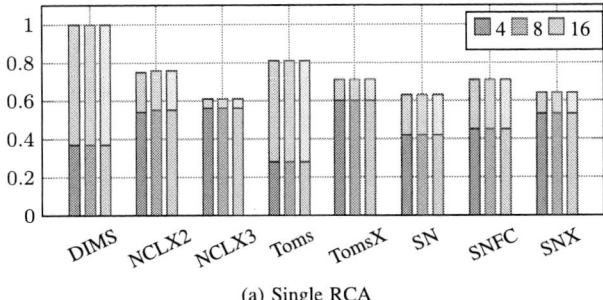

(a) Single RCA

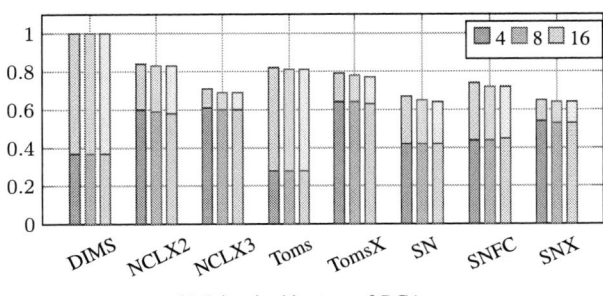

(b) 2-level adder tree of RCAs

Fig. 14. Area overhead of the presented circuits (normalized to the DIMS FA version)

- Sum delay Δ_S: The delay to generate the s output when c_{in} arrives, again assuming the same condition for a and b as before.

Given these parameters we can model the average delay Δ_{RCA} the RCA needs to produce its outputs. As already mentioned, the most important factor for the performance of a RCA is the length of its carry-chain. If the RCA consists of strongly indicating FAs, always the complete carry-chain has to be traversed, which means that their average case delay is equal to the worst case. Hence their overall delay (Eq. (1)) is given by the carry delay through $(n-1)$ FAs and the final delay to generate the MSB s signal or the c_{out} signal (depending on which circuit path is slower).

$$\Delta_{RCA}^{F} \approx (n-1)\Delta_L + \max(\Delta_L, \Delta_S) \qquad (1)$$

We only use the approximate symbol, because the equation does not fully consider the delay through the first FA of the chain. However for large n this discrepancy is not significant.

The calculation of the RCA delay for the case of weakly indicating FAs is a little bit more involved. In a first step we need to determine the average length of the longest carry-chain segment for a given input data width. Because we assumed uniformly distributed input values, for every possible input value combination there is an equal chance that a particular FA in the adder is able to generate its c_{out} signal with just the inputs a and b or that it has to wait for the c_{in} signal. This means that every possible input value configuration fits one of these 2^n equally likely cases. Hence we can simply enumerate every possible binary vector of length n, determine the longest sequence of zeros and take the average of these values. We denote this function with $l_c(n)$. A numerical analysis showed that $l_c(n)$ grows quite slowly ($l_c(8) = 2.1, l_c(16) = 3.2, l_c(24) = 3.8$). The final adder delay $\Delta_{RCA}^W(n)$ is then given by Eq. (2). Again the approximate symbol is used for the same reason as before.

$$\Delta_{RCA}^W(n) \approx l_c(n) * \Delta_L + \Delta_E + \Delta_S \qquad (2)$$

When comparing the two equations it is apparent that the weakly indicating FAs have quite a large performance advantage over the strongly indicating ones. Table II shows estimations for the delay parameters[6] for the weakly indicating FAs based solely on the gate delays involved (routing/wire delays are not considered).

TABLE II
ESTIMATED FULL ADDER DELAY PARAMETERS

Circuit	Δ_E	Δ_L	Δ_S
DIMS	8.5	4.75	3.75
NCLX2	3	2	2
SN	5.75	4	5.9
SNFC	3.75	2	6.6
SNX	3.75	2	3.5

As can be seen from the table, our proposed approaches, all perform better than the DIMS version. The SNX variant even has comparable delay to the NCLX2 version. However, these numbers must be interpreted cautiously, because circuit layout and the target technology also play an important role in the real-life performance of these circuits.

Another import point we need to address here is the delay when switching to the null phase. Again a RCA consisting of fully indicating FAs is at a disadvantage, because also in the null phase the carry signal has to traverse the complete carry-chain before every individual FA is reset. However, some of the weakly indicating circuits suffer from a similar albeit less severe issue. If the carry output of the SN FA was generated solely on the basis of the inputs a and b (i.e., for the case where $a = b$), then resetting just those inputs also resets c_{out}. If on the other hand the c_{in} signal was involved in generating c_{out} the circuits waits until c_{in} resets before letting c_{out} enter the null phase. Hence, the same carry segments that determined the delay of the circuit in the data phase also affect the null phase. Notice that the SNFC and SNX variants don't behave like that. A similar issue arises in the DIMS FA, where if $c_{in}.f$ is asserted, this rail must be deasserted before c_{out} can enter the null phase.

VII. CONCLUSION

We have analyzed QDI implementations of full adders with respect to peprformance and area overheads, with a focus on solutions that can be realized with modern standard libraries (extended by a 2-input C gate). As expected, strongly indicating approaches are generally slower, as they are bound to worst case processing time, unable to adapt to the data dependence of the latter. For weakly indicating solutions, in turn, a careful elimination of gate orphans is crucial.

---

[6]For this estimation, we assumed a unit delay of 1 for 2-input NAND/NOR gates, 1.25 for 3-input NAND/NOR gates, 0.75 for inverters, and 2 for C gates.

We have presented a novel, sorting network based implementation, both as a strongly indicating and two weakly indicating variants. Especially the latter turned out to have a very competitive processing speed, while their low area overhead makes them even more attractive.

In our evaluation we tried to be as generic and technology independent as possible through (a) using normalized values for area and delays for the comparisons, and (b) elaborating predictions for trends (over bit width). Furthermore we have explored different bit widths for the ripple-carry adder, as well as an adder tree.

REFERENCES

[1] M. Shams, J. C. Ebergen, and M. I. Elmasry, "Modeling and comparing CMOS implementations of the C-element," *IEEE Transactions on Very Large Scale Integration (VLSI) Systems*, vol. 6, no. 4, pp. 563–567, Dec 1998.

[2] P. A. Beerel, R. O. Ozdag, and M. Ferretti, *A Designer's Guide to Asynchronous VLSI*. Cambridge University Press, 2010.

[3] K. M. Fant, *Logically Determined Design – Clockless System Design with NULL Convention Logic™*. John Wiley & Sons, Ltd, 2005.

[4] A. N. Ismailoglu and M. Askar, "Application of Bit-level Pipelining to Delay Insensitive Null Convention Adders," in *2007 IEEE International Symposium on Circuits and Systems*, May 2007, pp. 3259–3262.

[5] X. Chang and Y. Lian, "A Quasi-Delay-Insensitive Dual-Rail Adder Working in Subthreshold Region," in *2009 IEEE International Symposium on Circuits and Systems*, May 2009, pp. 1569–1572.

[6] P. Balasubramanian and C. Dang, "A comparison of quasi-delay-insensitive asynchronous adder designs corresponding to return-to-zero and return-to-one handshaking," in *IEEE 60th International Midwest Symposium on Circuits and Systems (MWSCAS)*, Aug 2017, pp. 1192–1195.

[7] I. Obridko and R. Ginosar, "Minimal Energy Asynchronous Dynamic Adders," *IEEE Transactions on Very Large Scale Integration (VLSI) Systems*, vol. 14, no. 9, pp. 1043–1047, Sep. 2006.

[8] G. Paul, R. Reddy, C. R. Mandal, and B. B. Bhattacharya, "A bdd-based design of an area-power efficient asynchronous adder," in *2010 IEEE Computer Society Annual Symposium on VLSI*, July 2010, pp. 29–34.

[9] T. Verhoeff, "Delay-insensitive codes — an overview," *Distributed Computing*, vol. 3, no. 1, pp. 1–8, Mar 1988.

[10] R. Manohar and Y. Moses, "The Eventual C-Element Theorem for Delay-Insensitive Asynchronous Circuits," in *23rd IEEE International Symposium on Asynchronous Circuits and Systems (ASYNC)*, May 2017, pp. 102–109.

[11] P. McGee, M. Agyekum, M. Mohamed, and S. Nowick, "A Level-Encoded Transition Signaling Protocol for High-Throughput Asynchronous Global Communication," in *14th IEEE International Symposium on Asynchronous Circuits and Systems*, 2008, pp. 116–127.

[12] F. Huemer and A. Steininger, "Novel approaches for efficient delay-insensitive communication," *Journal of Low Power Electronics and Applications*, vol. 9, no. 2, 2019.

[13] J. Sparsø and J. Staunstrup, "Delay-insensitive multi-ring structures," *Integration, the VLSI Journal*, vol. 15, no. 3, pp. 313 – 340, 1993, special Issue on asynchronous systems.

[14] I. David, R. Ginosar, and M. Yoeli, "An efficient implementation of Boolean functions as self-timed circuits," *IEEE Transactions on Computers*, vol. 41, no. 1, pp. 2–11, 1992.

[15] A. Kondratyev and K. Lwin, "Design of Asynchronous Circuits by Synchronous CAD Tools," in *Proceedings 2002 Design Automation Conference (IEEE Cat. No.02CH37324)*, June 2002, pp. 411–414.

[16] J. Sparsø, *Introduction to Asynchronous Circuit Design*. DTU Compute, Technical University of Denmark, 2020, paperback edition available here: https://www.amazon.com/dp/B08BF2PFLN.

[17] C. Jeong and S. M. Nowick, "Optimization of Robust Asynchronous Circuits by Local Input Completeness Relaxation," in *2007 Asia and South Pacific Design Automation Conference*, Jan 2007, pp. 622–627.

[18] W. B. Toms, "Synthesis of quasi-delay-insensitive datapath circuits," Ph.D. dissertation, Department of Computer Science, University of Manchester, 2006.

Identification and Confinement of Fault Sensitivity Windows in QDI Logic

Florian Huemer, Robert Najvirt, and Andreas Steininger
Institute for Computer Engineering, TU Wien, Vienna, Austria
{fhuemer,rnajvirt,steininger}@ecs.tuwien.ac.at

Abstract—Asynchronous, specifically QDI circuits are known to exhibit high resilience against faults affecting the timing. At the same time, their event based operation principle makes them susceptible to glitches. While synchronous circuits obtain high resilience through temporal masking that is established through the sampling of data by flip flops, asynchronous designs, by trying to be flexible about the phases of data validity, often have lower masking capabilities. Consequently, previous work has proposed to narrow down the windows in which data changes are accepted, in order to improve the temporal masking in QDI designs.

In this paper we study the natural resilience of different QDI templates in more detail and quantitatively determine the windows of vulnerability through extensive fault injection experiments in simulation. To this end we propose a novel way of visualizing and analyzing the sensitivity windows that aids in identifying the key dependencies and vulnerabilities. In addition we introduce and evaluate two low-cost extensions for the pipeline registers which allow either to stall the operation in case a glitch creates an illegal symbol, or to prevent the creation of an illegal symbol in the first place.

I. INTRODUCTION

In contrast to synchronous circuits with their rigid time-driven operation dictated by the clock, asynchronous circuits, in particular Quasi Delay Insensitive (QDI) circuits, employ a closed loop control, established by handshake protocols, to adapt their speed of operation to the respective operating conditions. This makes their timing very robust and hence initially qualifies them for building resilient circuits. However, to leverage that potential their behavior with respect to faults in the value domain must be understood and optimized as well. In synchronous circuits it is evident how flip flops provide temporal masking and thus make the circuit immune against transient faults during certain phases. QDI circuits, in contrast, are deemed susceptible to transient faults in the value domain, due to their transition-centric operation. Still, however, these circuits also benefit from immanent masking effects, but these are more difficult to identify, as they depend on implementation, pipeline fill level, path delays etc. For any experimental assessment by physical or simulated fault injection this makes the parameter space to be covered huge. In addition, rather than just giving one or several key statistical descriptors for the masking – like the size of the sensitive window relative to the

This research was partially supported by the project ENROL (grant I 3485-N31) of the Austrian Science Fund (FWF).

clock period in a synchronous design – it becomes important to comprehend the relevant dependencies.

One contribution of this paper is a novel method for visualizing fault sensitive windows that gives detailed information about the sensitivity of individual signals and its dependence on pipeline fill level in a compact and intuitive way. A second contribution is a thorough analysis of a Weak-Conditioned Half Buffer (WCHB) based pipeline, along with an apples-to-apples comparison of different methods proposed in literature for enhancing its resilience, first in theory and then based on our proposed experimental approach that is based on fault injection simulations. Finally, as a third contribution, we present two novel schemes that prevent the generation/propagation of illegal codes, as they often result from faults in QDI circuits, across (WCHB) pipeline stages, and thus contribute to enhancing resilience in the value domain.

The paper is structured as follows: The next section gives a brief introduction into the principles of asynchronous circuits. Then Sec. III reviews existing literature about enhancing and assessing their resilience. Sec. IV presents the design templates and mechanisms we consider in our comparison and analyzes their resilience from a theoretical point of view, followed by an experimental analysis in Sec. V. Finally Sec. VI introduces our novel enhancement schemes and assesses their effectiveness, before Sec. VII concludes the paper.

II. BACKGROUND

Where synchronous circuits use the clock signal to control data transfer between storage elements (e.g., pipeline stages), asynchronous circuits use some form of closed-loop handshaking protocol. This handshake mechanism (usually) involves two signals, called request (req) and acknowledgment (ack). The data source uses the req signal to indicate the availability of new data to the sink, which then acknowledges the reception using the ack signal[1]. Depending on the number of transitions on these two wires for one complete handshaking cycle, the protocol can be classified as 2-phase or 4-phase. A 4-phase handshake is defined by the switching sequence $req\uparrow$, $ack\uparrow$, $req\downarrow$, $ack\downarrow$, where the arrow symbols denote rising and falling transitions, respectively. Hence, the handshake signals always start and end the handshaking cycle with the same logic value. In contrast to that, 2-phase protocols only use two transitions and leave the handshake signals in the opposite logic state.

[1]This explanation assumes push channels, pull channels will not be considered in this work.

978-1-7281-8494-4/20 $31.00 © 2020 IEEE

This means that e.g., $req\uparrow, ack\uparrow$ or $req\downarrow, ack\downarrow$ both constitute complete 2-phase handshakes.

The described request mechanism does not need to be implemented as a dedicated req wire. It is also possible to implicitly encode the request into the transmitted data using some DI code [1], which leads to the class of QDI circuits [2]. The sink then has to use a completion detector (CD) to decide when the received data is complete (i.e., valid) and can thus be consumed and acknowledged. QDI circuits can also be implemented using 2-phase or 4-phase protocols. However, for circuits that actually process data (in contrast to just transmitting it, e.g., [3]), practically only the 4-phase variant is relevant. Hence, in this paper we only focus on this style. For the same reason we will only consider the dual-rail (DR) data encoding as Delay-Insensitive (DI) code. The DR code uses two rails (i.e., wires) for each bit, the true and false rail. For a DR bit d, we denote these rails by $d.t$ for the true rail and $d.f$ for the false rail. In every handshaking cycle all data rails start out in a low state (spacer) and only one rail of each DR bit can go high. After acknowledgment all rails switch back to low and the whole process start over. Hence the receiver can simply use an OR gate to check for completion on each received DR bit. The individual phases of this protocol are also referred to as *null* and *data* phase.

Fig. 1 shows an example timing diagram of the transmission of two DR bits d_0 and d_1. The bit order of the transmitted bits in the figure is given by $(d_1.t, d_1.f, d_0.t, d_0.f)$. The transmitted binary data (d_1, d_0) in the figure is given by $(0, 0)$ followed by $(1, 0)$.

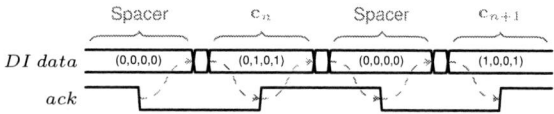

Fig. 1. 4-phase Protocol

It is important to stress that the order in which the transitions on the individual data rails arrive at the receiver does not matter. Furthermore, the individual gate and wire delays in a circuit can vary arbitrarily and the circuit is still guaranteed to work correctly. The only restriction on the delays is the so called isochronic fork constraint, which demands that for certain wire forks the delays of the individual paths after the fork must be equal. This is also the reason why this class of circuits is referred to as "only" *quasi* DI.

This fundamental property of QDI circuits, with respect to their delays, constitutes the key for obtaining tolerance against PVT variations. The disadvantage that is associated with this robustness is that a QDI circuit always entails a high area overhead when compared to corresponding synchronous or asynchronous bundled data circuits. The reason for this area overhead is the additional cost associated with the DI code, the required CDs and the more complex logic gates.

III. Related Work

Our analysis will focus on transient faults, as these are much more frequently encountered during different missions,

especially in the shape of single event transients (SETs), i.e., short voltage pulses caused by radiation particle hits. For synchronous designs, there is an abundance of literature available for mitigating SETs and avoiding single event upsets (SEUs) they cause in storage elements. These concepts cannot be directly applied to asynchronous circuits for two reasons: (i) The effects of SETs are different in asynchronous logic, and (ii) most of the synchronous solutions rely on synchrony assumptions between replica, which do not (naturally) hold in asynchronous architectures [4]. In order to cope with (i), an appropriate modeling of faults (propagation and masking) in asynchronous logic is essential, which again differs from the synchronous models like [5] or [6]. In an early approach [7] built an automated verifier for speed independent circuits based on trace theory, which is a very natural means for describing the sequences of transitions relevant for the operation in asynchronous circuits. Later, [8] also used trace theory (among others) for modeling the correct (and incorrect) behavior of asynchronous circuits, more specifically interfaces. In [9] the authors perform a very thorough analysis of SET effects in a selected QDI function block. Their focus is first on the gate level and then on channels (with 4-phase/RTZ protocol), and they use handshaking expansion notation (HSE) to describe the circuit behavior. They identify deadlock, synchronization failure, token generation and token consumption as possible SET effects. In [10] a similar investigation has been conducted, again by means of HSE notation. Here, the treatment of glitch effects and their coverage is more formal, but also a validation through transistor level simulation is given. In a relatively informal case distinction [11] analyzes the consequences of SETs in 4-phase QDI circuits, and concludes that possible effects are blocking, filtering, delay fault and soft error. Based on this analysis three hardening techniques are elaborated. By means of signal transition graphs (STGs) [12] performs a very comprehensive study of SEU (rather than SET) effects in QDI circuits, while later [13] also uses STGs to investigate SETs in QDI network-on-chip links. They propose a bundle of techniques to protect QDI communication links and interfaces from glitches. Also with a focus on SEUs [14] applies symbolic simulation for an exhaustive coverage of all possible behaviors under faults. A more experimental approach is pursued in [15]. Here a simulated gate level fault injection study is performed to compare the sensitivity of synchronous versus asynchronous logic blocks to transient faults and to analyze the respective masking effects. In almost all these approaches the assessment, if any, is limited to a study of overheads. On the other hand, in those few publications where actual fault injection into asynchronous logic has actually been performed (in simulation or hardware), the focus is very narrow [10], or those experiments serve a purpose other than supporting a model [16]–[18]. This makes it difficult to perform an apples-to-apples comparison of the different resilience enhancement approaches proposed.

Fundamentally, the adverse effect of particle hits can be largely mitigated by the use of specific rad hard technologies. In their simplest form they provide just a design kit and library

978-1-7281-8494-4/20 $31.00 © 2020 IEEE

with larger feature sizes, which results in higher capacitances and higher critical charge, such that the charge induced by a particle hit does not cause voltage pulses with considerable amplitude [19]. Consequently, such libraries cannot leverage the benefits of technology scaling, and hence their overheads become progressively higher. A more scalable concept among the measures for obtaining SET tolerance is, like in the synchronous domain, full or partial [20] replication. There are, however, two important differences: (a) Concurrent voting is not straight¬forward, as each instance is running at its own, self-timed, pace [4]; but (b) the control loop established by the handshake allows to keep the faster instance waiting until the slower has its result available as well. This fundamental principle is outlined in [12] where a C gate performs this synchronization task. This principle has been often adopted and extended, like in [11], [18]. There are, however, two fundamental problems: (1) Without further measures (that ultimately imply undesired timing assumptions) a non-responding instance (omission fault) can deadlock the whole redundant architecture. (2) A short SET on the slower one of the redundant paths may lead to early completion and ultimately cause desynchronization of the replica. Beyond a mere replication of function blocks, a clever mutual synchronization of similar functions (like such applied to neighboring bits in a word) or the reduction of the redundant function block to a lean synchronization component are proposed (e.g., in [11]). In [21] the authors use a combination of spatial redundancy and guard gate to harden a controller against radiation. Detailed knowledge about (legal and illegal) protocol states or tokens has often been leveraged for error detection in internal and external asynchronous links, like in [4], [10], [13], [22], [23] and [24]. In [25] information redundancy is leveraged to make a whole processor low power and asynchronous at the same time. In particular, [26] relies on checking for an illegal state of a bit's DR representation before actually latching any of the two rails, while the work in [27] is based on replication of the cell plus DR conversion. A complete radiation-hard-by-design QDI processor (DD1) has been presented in [18]. To achieve SET tolerance it essentially relies on duplication and cross coupling, as well as memory protection techniques.

IV. QDI DESIGN STYLES

For our analysis we primarily look into WCHB-based design styles and some derivatives. Fig. 2 shows a single-bit WCHB pipeline with three stages. The storage elements in this buffer are C gates[2]

The operation principle of this circuit is quite simple. Assume the input data rails and the input acknowledgment (ack_{in}) are zero, i.e., the circuit is in the null phase. Due to the inverter the enable signals (en) are high, which arms the C gates for rising transitions on the data rails. After some input data arrives and one of the C gates (in the first

---

[2]The Muller C-element, or short C gate, is a fundamental gate in asynchronous logic. Its function is to output the logic level seen at its inputs when these match, and to retain the last valid output state otherwise. It can hence also be viewed as an AND gate with hysteresis.

buffer) actually switches to one, the CD, i.e., the OR gate, will eventually generate the output acknowledgment ack_{out}. At the same time this transition also travels through the whole pipeline, setting the respective C gates in each stage, until the data appears at the output of the pipeline. Since the input data has been acknowledged by the fist stage, the input rails may now enter the null phase again. This null phase will then propagate through the pipeline as well. However, it will only be able to reset the last stage in the pipeline if the output data is acknowledged by a falling edge on ack_{in} first. This means a single WCHB will keep its stored value (or spacer) until the succeeding pipeline stage acknowledges the data (or spacer), by toggling the acknowledge input of the respective stage (using its CD).

When implementing n-bit WCHBs, completion detection is performed individually on each DR bit by an OR gate. The outputs of these OR gates are then merged by an n-input C gate to generate the output acknowledgment ack_{out}

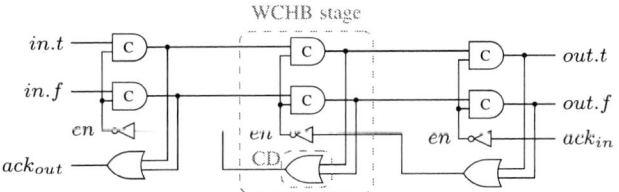

Fig. 2. 3-stage single bit WCHB pipeline

A. Fault Effects and Sensitivity Windows

Fig. 3 shows a timing diagram of a multi-bit WCHB over a single handshake cycle. The shown data signals correspond to the data outputs of the buffer, i.e., the outputs of the C gates. The DR bits d_0 and d_n symbolize the pair of data signals with the largest skew between them (windows B-C and E-F).

As soon as the input acknowledgment ack_{in} goes low (A) a WCHB waits for the data-phase and thus all its C gates are armed for rising transitions. The C gates are only disabled when the next stage acknowledges the received data (D). This leaves quite a large time window (A-D) where the buffer is susceptible to faulty (rising) input transitions. We say that the buffer is now in a (fault-)accumulating state, since all input transitions (faulty and valid ones alike) will be captured into the C gates. This is in stark contrast to e.g., synchronous designs, where there is only a single point in time (i.e., the clock edge) where data is captured by the storage elements.

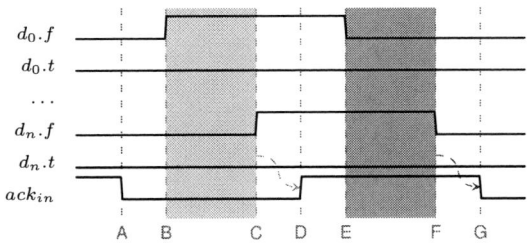

Fig. 3. WCHB time windows

Depending on when exactly a fault strikes a data wire during the data phase and erroneously sets a C gate, one can observe vastly different effects. If a fault strikes a data wire that should transition anyway, in the best case only the point in time when this transition happens is shifted, which is completely tolerable in the context of QDI circuits. However, given sufficient skew between the data rails, in the worst case such a fault can deadlock the whole circuit. Striking a data rail that should remain zero a fault can (besides the deadlock) cause an invalid output pattern for a DR bit (i.e., both rails set to one) or a valid but incorrect data token, depending on whether the valid transitions on the respective other DR signal arrives before the buffer is closed.

After ack_{in} has been asserted (D) to acknowledge the received data the C gates in the buffer are again armed for falling input transitions. Hence there is a time window (D-G) where the buffer is susceptible to faulty (falling) input transitions. However, since the null phase does not carry any actual data, faults don't affect the transmitted data but can again lead to deadlocks.

Since the input acknowledgment signal ack_{in} basically represents the enable signal for all the C gates in a buffer, faults affecting this signal can have severe consequences on a circuits. These range from deadlocks to lost data items or illegal output patterns. The ack_{in} (or en) signal is vulnerable during phases where the inputs to the C gates differ in their logical value (i.e., the C gate is in state-holding mode). In such a situation a single input change can flip the value of the C gate. As Fig. 3 shows, at time (C) the output of the buffer is complete, hence eventually the output acknowledgment ack_{out} would be asserted, which in turn leads to the deassertion of all input data rails. However the output of the buffer must not enter the null phase until ack_{in} is asserted, which means that there is a sensitive window for the acknowledgment signal between (C) and (D). A similar situation arises during the null phase between the points (F) and (G).

We see that although in general delays are not relevant for the correct behavior of QDI circuits, the exact fault behavior of such a circuit is highly dependent on the input skew, the involved circuit delays, the speed of operation and the actual data that is being processed. This means that the sensitivity of a buffer cannot be evaluated completely statically but must incorporate environmental conditions as well as concrete circuit details.

B. Fault Mitigation Strategies

To address these fault sensitivity windows inherent to the WCHB discussed in the previous section, several strategies have been proposed in literature. Here, we present a selection of these approaches, which will also be analyzed in more detail in the following section.

Most WCHB modifications aim at shortening the time windows in which the buffer stores input transitions. One way to achieve this, is to use two CDs for every buffer, one for the input data and another one for the output data. Consequently we refer to this buffer as the *Dual CD* WCHB. The C gates of

the buffer are then only armed when there is actually data at the input (point (C) in Fig. 3). This idea is presented in more detail in [13], where it is referred to as *normally closed latch*.

Moreover, [13] further proposes another slightly different approach, which uses asymmetric C gates as storage elements for the buffer. These C gates have one additional (asymmetric) input that must be asserted in order to set the C gate, but does not need to be deasserted to reset it again, like a normal input would have to be. The asymmetric input is then fed by the inverted output of the buffer's CD, which ensures that the C gates are disarmed as soon all input transitions arrived, effectively closing the sensitivity window at point (C) in Fig. 3. In case of the DR code this can be done on a per-bit basis by using the output of the individual OR gates, which has the benefit of closing the C gate of the respective other rails as soon as one transition arrived. This approach prevents the capturing of the invalid DR state (11). However, blindly capturing the first transition creates a potential of forwarding a wrong value (50% when assuming random faults). In that sense it operates similarly to a Mutex, with the important difference that depending on the timing of the feedback path, there may still be cases where both outputs are set simultaneously. This buffer modification can be beneficial in combination with an error correcting code (ECC) on top of the DR code: It prevents the protocol from being upset by an illegal DR state, while the potential corruption of the data value it causes can be undone by the ECC. In this paper we refer to this approach as the *Locking* WCHB.

Another approach to avoid the lasting consequence of a faulty transition is to use different storage elements. For that purpose a D latch based Mousetrap-style pipeline structure as proposed in [3] can be used. Here a simple buffer control circuit, consisting of just a single XOR gate, is responsible to enable and disable the buffer's D latches. When the D latches are transparent, input glitches caused by SETs can freely propagate thorough the latch to the output. Unless the latch is closed in exactly this time instance, the latch will not store the faulty value. Similar to the previous approach the buffer is closed as soon as the CD detects the data phase. Although strictly speaking this circuit is not QDI because it introduces a small timing constraint, we still want to include it in our survey since it should show quite a different behavior in the analysis and will refer to it as MTDHB (Mousetrap-style D Latch half buffer). Note, however that this buffer has not been proposed to improve fault-tolerance.

Finally, we also want to mention a completely different approach proposed by Jang and Martin in [12]. This is not really a design or buffer style on its own, rather a technique for hardening existing QDI designs. Here the original design is duplicated and both copies are interlocked with C gates that essentially vote on every intermediate signal of the circuit. The authors formally show that this scheme is able to tolerate (single) faults on any internal signal, which means that in our analysis it should not show any erroneous behavior. However, it is easy to see that this approach entails more than double the area overhead of the original design. Moreover, the additional logic for synchronization of the two replicas also results in a slightly slower circuit. Following the terminology used in [12] we refer to WCHB using the described technique as a *doubled-up double-checking* (DD) WCHB.

V. EXPERIMENTAL ANALYSIS

To experimentally analyze what fault sensitive windows the different QDI buffer styles have, we use a Questa based simulation to inject SETs. Fig. 4 shows our fault-injection simulation target: a 4 stage, 2 bit, DR QDI pipeline connected to a data generator at its input and a checker at its output. The pipeline was generated for each buffer style under evaluation and uses inertial gate delays and no wire delays. The 4 data rails between buffers 1 and 2 as well as the acknowledgment input to buffer 2 were chosen as victim wires. Whether an injected fault had an effect on the circuit was checked in both the data generator and checker, which compared timing and values to a reference run, as well as checking for protocol and coding violations. Whenever a deviation from the undisturbed *golden* run was detected, the behavior was classified as follows:

- Timing Deviation: A transition happened earlier or later than expected. The circuit being DI, this is not a fault, but rather an observation.
- Value Fault: A wrong data value was delivered to the output.
- Code Fault: An invalid DI code word was observed at the output (i.e., both rails of a DR bit high).
- Glitch: A signal changed its value twice during a protocol phase. This includes protocol violations (e.g., acknowledgment before data completion)
- Deadlock: The circuit reached a state where no further transitions were possible.

For the depiction of the results, we consider these classes to be ordered in ascending importance – whenever a fault injection triggered faults from different classes, only the one with the higher importance was plotted and counted towards the results. An example would be a DR output expected to change from the null phase (00) to a logical 1 (10) which, as a result of a fault injection changes from (00) to a logical 0 (01) at an unexpected time, generating both a value fault and a timing deviation event in the checker. If subsequently the expected transition arrived causing a code fault (11) we would consider the fault injection to yield a code fault and disregard the value fault and timing deviation.

It is important to note that the effects of a fault injection were only observed at the outputs of the pipeline with one pipeline stage between the victim wires and the checker. The motivation behind this choice was to let the additional pipeline

stages perform logical and temporal masking as it would occur in a normal pipeline. Attaching the strict checker directly to the buffer that was targeted in our fault injection would show unnecessarily high fault rates when the often less strict pipeline stage simply does not let many faults through to the checker at the output. While allowing to get better results for buffer styles that propagate fewer faults, this topology also hides some internals that would allow us to differentiate even better between the fault manifestations. When a data word is erroneously removed from the pipeline due to the injected fault, the checker would probably only see a timing deviation and a value fault for some of the following words extracted from the pipeline. The same holds true for spurious data items injected into the pipeline during the null phase.

Fig. 5a shows a simulation trace for the original WCHB. First, the simulation started with a pre-run, where it was counting several rising transitions on the ack_{in} wire to skip the initial phase, where the pipeline only starts to fill up, which is not visible in the figure. Afterwards, 400 ps pulses were injected into the victim wires in 250 ps steps between 0 ns and 120 ns from the end of the pre-run (marked by the 0 ns point on the x-axis). Markers indicate injection times of pulses that had an observable effect on the pipeline, their colors representing the classification of that effect.

As we have seen from the analysis in the previous section, the timing of the input signals to a pipeline has a high impact on the fault sensitivity windows. By operating the circuit at a specific speed (i.e., handshake rate) the external signals determine how much time the circuit spends in the different protocol phases and hence in its sensitive windows. This is again in stark contrast to a synchronous designs, where the input signals simply don't have that much "power" over the circuit. Hence, a single simulation trace (Fig. 5a) alone only yields very little information about fault behavior of a circuit, since it only shows one specific operation point. For that reason, an essential part of the simulation setup was the possibility to choose delays of the source and sink when generating new input words and acknowledgments respectively. It allowed us running the fault injection simulation of the same pipeline with a variety of timing settings, gradually changing its operation from token-limited (where the pipeline stages mostly wait for valid data words to arrive) to bubble-limited (where the pipeline stages receive valid data at their inputs, but need to wait for the acknowledgment from the succeeding stage before being allowed to store the new data word).

Fig. 5b shows the results of such a timing variation for the same WCHB pipeline simulated in Fig. 5a: For each signal, abutted horizontal stripes represent the results of 11 different simulation runs, in which the pipeline transitioned from bubble-limited (top) to token-limited (bottom) operation. A stripe is colored blue where the signal is low and orange where it is high. Note that Fig. 5a shows the topmost WCHB simulation stripe from Fig. 5b. In the same way the other subgraphs in Fig. 5 illustrate the behavior of the pipeline using the alternative pipeline implementation styles presented in Sec. IV. This representation style, that allows to pack a large amount of

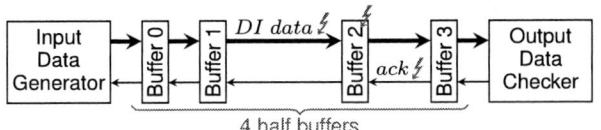

Fig. 4. Simulation Setup

2020 Austrochip Workshop on Microelectronics

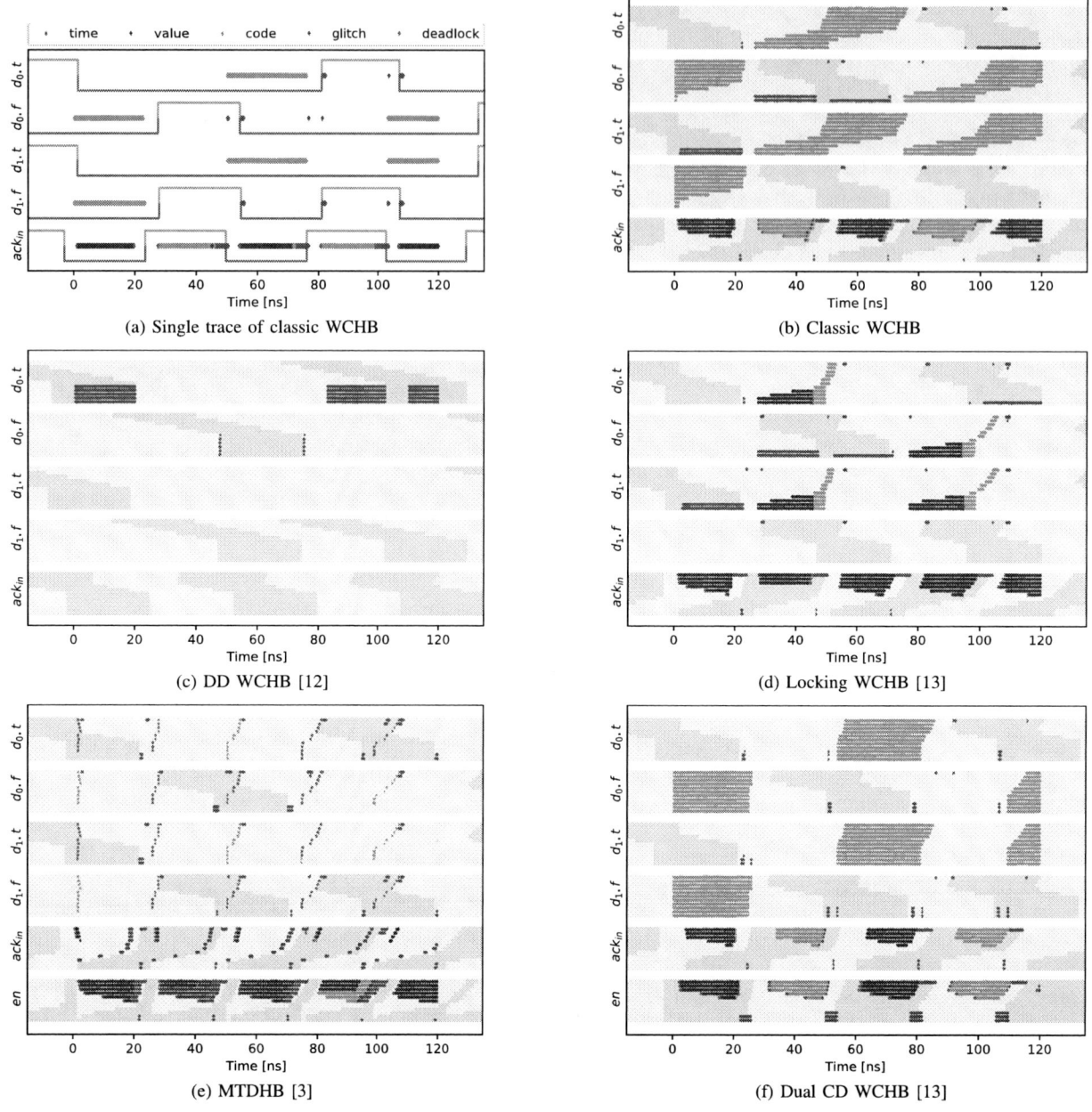

Fig. 5. Simulation Results

information about the fault behavior of a buffer into a single figure, is one of our key contributions of this work.

It allows us to see how the external interface timing affects the sensitivity to faults of the different pipeline implementations when subject to SETs. The data rails for the classic WCHB implementation show how the inactive rail is sensitive to produce a code fault the entire time the receiving C gate is accumulating, irrespective of whether the pipeline runs token- or bubble-limited. The Locking WCHB significantly reduces the sensitive windows by correctly preventing code faults in bubble-limited operation after a transition on one of the two rails was captured by a C gate. It only fails to prevent code

faults for a short time corresponding to the feedback delay for locking. In token-limited operation, the injected pulse is captured and the correct and expected transition on the other data rail is prevented from turning the valid, albeit incorrect, value into a code fault.

Unsurprisingly, the DD WCHB style proves to be insensitive to SETs in all operation modes whereas the Dual CD WCHB style brings little to no improvement to the sensitivity windows. The MTDHB shows very narrow sensitivity windows on the data rails while the enable signal (the signal that activated the D latches of the buffer) is sensitive most of the time. Faults on this signal also have a wide range of possible effects.

978-1-7281-8494-4/20 $31.00 © 2020 IEEE   34

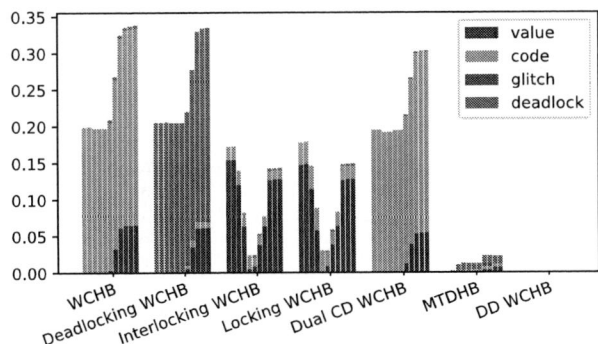

Fig. 6. The number of fault injection simulations with an observable effect on the pipeline

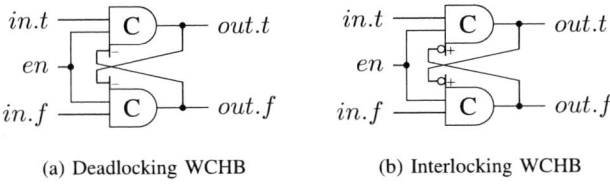

(a) Deadlocking WCHB   (b) Interlocking WCHB

Fig. 7. Proposed WCHB modifications

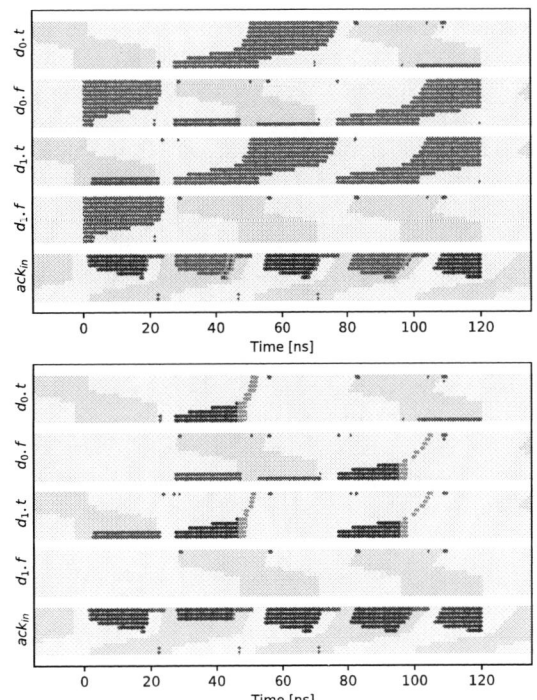

Fig. 8. Simulation results for the proposed buffer (top: Deadlocking WCHB, bottom: Interlocking WCHB)

Fig. 6 shows the ratio of injected faults that had an observable effect other than a timing deviation to all injected faults. Note that buffer styles which will be introduced in the following section are also included in this figure. To make a fair comparison and prevent speed differences from influencing the results, faults are only considered during one handshake cycle between two rising edges of the acknowledgment wire. For each considered buffer style, the 11 bars show the results for the 11 simulations depicted in Fig. 5 and Fig. 8. It is apparent that the robustness of the simulated pipelines clearly depends on the external timing. By observing a sweep of simulations with varying timing, one can qualitatively assess the effectiveness of fault mitigation techniques like the locking WCHB design instead of barely looking at numbers without the knowledge of whether the circuit was simulated with token- or bubble-limited timing. It could even happen, that a developed fault mitigation technique which is not effective in practice might appear to be advantageous in fault injection simulations, where in fact, additional delay due to added logic gates changed the circuit operation from token-limited to balanced which in itself could bring a significant improvement of fault sensitivity windows. The figure also shows that, depending on the operation mode a circuit is actually used in, it may not pay off to invest in very high overhead fault mitigation strategies, because simple and comparatively cheap approaches also yield quite good robustness.

VI. PROPOSED BUFFER ENHANCEMENTS

Fig. 7 shows two half-buffer designs (with comparatively low hardware overhead) that try to mitigate the fault-accumulating behavior of the classical WCHB discussed in Sec. IV-A. Both use cross-coupled asymmetric C gates, whose outputs are fed back to the asymmetric input of the respective other gate.

For the *deadlocking WCHB* the feedback inhibits the buffer from entering the null phase if erroneously both C gates are set, effectively causing a deadlock and preventing the circuit from processing possibly faulty data. This can be useful in applications where the correctness of the output is crucial, while deadlocking is not harmful (fail stop). The *interlocking WCHB* only allows the first transition at its input to propagate, thus prohibiting the invalid DR (11). In this sense it is similar to the locking WCHB design proposed in [13] and discussed in Sec. IV-B. However, one important difference is that their approach uses the output of the CD to deactivate the corresponding C gates (i.e., prohibiting them from switching to one) in the buffer. While this allows the use of arbitrary DI codes, it also prolongs the feedback path by the delay of the CD, which keeps the buffer open and thus susceptible to SETs on its inputs for a longer time window. Another more subtle difference to the approach in [13] is that in order to prevent an erroneous input transition from setting a C gate, after completion *all* C gates are switched to a state-holding mode (i.e., the output is driven by the internal storage loop). In our approach only the C gate connected to the rail that did not transition to high is switched to the state-holding mode (keeping its zero value).

Fig. 8 and Fig. 6 show the results of the analysis. It is clearly visible how the deadlocking WCHB, as expected, turns all code faults seen on the data rails in the classic WCHB into deadlocks, albeit without changing the sensitive window. The interlocking WCHB in turn significantly shortens the sensitive windows, since in the bubble limited case the correct transition appears early, and afterwards the interlocking closes the sensitive window. The non-zero size of the remaining window is due to the propagation delay for the locking to become active. Note that these windows are slightly shorter than those found for the approach from [13] in the previous section, due to the shorter feedback path. In the token limited case, there is a potential for faults on the non-switching rail to arrive before the correct transition on the other rail and thus lock the buffer in an incorrect (but valid) state. This is indicated by the windows with value faults that match the size of the sensitive windows in the original WCHB.

VII. CONCLUSION

We have analyzed the sensitivity of different variations of a WCHB to transient faults. Given that this sensitivity strongly depends on the speed of source and sink, our conclusion was that a systematic analysis requires a visualization of this dependence. Our proposed solution here is a graphical representation of the sensitive windows for each relevant signal, aggregated for different settings of source speed and sink speed, and showing a color code for the observed effect of a fault at the point corresponding to its injection. In another type of visualization we have compiled the absolute number of observed effects summed up over all relevant signals of a given buffer implementation, again plotted as a trend over various settings of sink and source speed (i.e. going from bubble limited to token limited). The required data for these plots are generated in extensive fault injection experiments into a simulation model. Using these compact representations we have identified the key vulnerability of the classical WCHB and proposed two enhancements, namely an interlocking buffer and a deadlocking buffer. For both we have sketched a target use case and given evidence for their proper operation through our graphical analysis.

On the long run, our approach aims at elaborating a generic model of the sensitivity of a QDI block to SETs which allows to correlate the types and rate of effects observed with design parameters like pipeline style, protocol, latch implementation, activity profile and others. Based on this knowledge, the assessment of a specific circuit's susceptibility to SETs should be relatively straightforward.

REFERENCES

[1] T. Verhoeff, "Delay-insensitive codes — an overview," *Distributed Computing*, vol. 3, no. 1, pp. 1–8, Mar 1988.

[2] A. J. Martin, *The Limitations to Delay-Insensitivity in Asynchronous Circuits*. New York, NY: Springer New York, 1990, pp. 302–311.

[3] P. McGee, M. Agyekum, M. Mohamed, and S. Nowick, "A Level-Encoded Transition Signaling Protocol for High-Throughput Asynchronous Global Communication," in *14th IEEE International Symposium on Asynchronous Circuits and Systems*, 2008, pp. 116–127.

[4] T. Verdel and Y. Makris, "Duplication-based concurrent error detection in asynchronous circuits: shortcomings and remedies," in *Proceedings of the 17th IEEE International Symposium on Defect and Fault Tolerance in VLSI Systems (DFT)*, 2002, pp. 345–353.

[5] Y. S. Dhillon, A. U. Diril, and A. Chatterjee, "Soft-error tolerance analysis and optimization of nanometer circuits," in *Design, Automation and Test in Europe*, March 2005, pp. 288–293 Vol. 1.

[6] P. Liden, P. Dahlgren, R. Johansson, and J. Karlsson, "On latching probability of particle induced transients in combinational networks," in *Proceedings of IEEE 24th International Symposium on Fault- Tolerant Computing*, June 1994, pp. 340–349.

[7] D. L. Dill, "Trace theory for systematic verification of speed-independent circuits," Ph.D. dissertation, Carnegy Mellon University, 1988.

[8] A. Yakovlev, "Structural technique for fault-masking in asynchronous interfaces," *IEE Proceedings E - Computers and Digital Techniques*, vol. 140, no. 2, pp. 81–91, March 1993.

[9] C. LaFrieda and R. Manohar, "Fault detection and isolation techniques for quasi delay-insensitive circuits," in *International Conference on Dependable Systems and Networks, 2004*, June 2004, pp. 41–50.

[10] S. Peng and R. Manohar, "Efficient failure detection in pipelined asynchronous circuits," in *20th IEEE International Symposium on Defect and Fault Tolerance in VLSI Systems (DFT'05)*, 2005, pp. 484–493.

[11] Y. Monnet, M. Renaudin, and R. Leveugle, "Hardening techniques against transient faults for asynchronous circuits," in *11th IEEE International On-Line Testing Symposium*, July 2005, pp. 129–134.

[12] W. Jang and A. J. Martin, "SEU-tolerant QDI circuits [quasi delay-insensitive asynchronous circuits]," in *11th IEEE International Symposium on Asynchronous Circuits and Systems*, March 2005, pp. 156–165.

[13] W. J. Bainbridge and S. J. Salisbury, "Glitch Sensitivity and Defense of Quasi Delay-Insensitive Network-on-Chip Links," in *15th IEEE Symposium on Asynchronous Circuits and Systems*, May 2009, pp. 35–44.

[14] Y. Monnet, M. Renaudin, and R. Leveugle, "Formal Analysis of Quasi Delay Insensitive Circuits Behavior in the Presence of SEUs," in *13th IEEE International On-Line Testing Symposium*, July 2007, pp. 113–120.

[15] R. P. Bastos, Y. Monnet, G. Sicard, F. Kastensmidt, M. Renaudin, and R. Reis, "Comparing transient-fault effects on synchronous and on asynchronous circuits," in *15th IEEE International On-Line Testing Symposium*, June 2009, pp. 29–34.

[16] Y. Monnet, M. Renaudin, and R. Leveugle, "Asynchronous circuits sensitivity to fault injection," in *10th IEEE International On-Line Testing Symposium*, July 2004, pp. 121–126.

[17] B. Rahbaran and A. Steininger, "Is asynchronous logic more robust than synchronous logic?" *IEEE Transactions on Dependable and Secure Computing*, vol. 6, no. 4, pp. 282–294, Oct 2009.

[18] S. Keller, A. J. Martin, and C. Moore, "DD1: A QDI, Radiation-Hard-by-Design, Near-Threshold 18uW/MIPS Microcontroller in 40nm Bulk CMOS," in *21st IEEE International Symposium on Asynchronous Circuits and Systems*, May 2015, pp. 37–44.

[19] D. J. Barnhart, T. Vladimirova, M. N. Sweeting, and K. S. Stevens, "Radiation hardening by design of asynchronous logic for hostile environments," *IEEE Journal of Solid-State Circuits*, vol. 44, no. 5, pp. 1617–1628, May 2009.

[20] F. A. Kuentzer and M. Krstic, "Soft Error Detection and Correction Architecture for Asynchronous Bundled Data Designs," *IEEE Transactions on Circuits and Systems I: Regular Papers*, pp. 1–12, 2020.

[21] F. A. Kuentzer, M. Herrera, O. Schrape, P. A. Beerel, and M. Krstic, "Radiation Hardened Click Controllers for Soft Error Resilient Asynchronous Architectures," in *26th IEEE International Symposium on Asynchronous Circuits and Systems (ASYNC)*, May 2020, pp. 78–85.

[22] Y. Shi and S. B. Furber, "Error checking and resetting mechanisms for asynchronous interconnect," in *Proceedings of the 18th UK Asynchronous Forum, Newcastle University*, 2006, p. 4.

[23] Y. Shi, S. Furber, J. Garside, and L. Plana, "Fault Tolerant Delay Insensitive Inter-chip Communication," in *15th IEEE Symposium on Asynchronous Circuits and Systems*, May 2009, pp. 77–84.

[24] W. Kuang, P. Zhao, J. S. Yuan, and R. F. DeMara, "Design of Asynchronous Circuits for High Soft Error Tolerance in Deep Submicrometer CMOS Circuits," *IEEE Transactions on Very Large Scale Integration (VLSI) Systems*, vol. 18, no. 3, pp. 410–422, March 2010.

[25] M. Marshall and G. Russell, "A Low Power Information Redundant Concurrent Error Detecting Asynchronous Processor," in *10th Euromicro Conference on Digital System Design Architectures, Methods and Tools (DSD 2007)*, Aug 2007, pp. 649–656.

[26] K. T. Gardiner, A. Yakovlev, and A. Bystrov, "A C-element Latch Scheme with Increased Transient Fault Tolerance for Asynchronous Circuits," in *13th IEEE International On-Line Testing Symposium (IOLTS 2007)*, July 2007, pp. 223–230.

[27] N. Julai, A. Yakovlev, and A. Bystrov, "Error detection and correction of single event upset (SEU) tolerant latch," in *IEEE 18th International On-Line Testing Symposium (IOLTS)*, June 2012, pp. 1–6.

978-1-7281-8494-4/20 $31.00 © 2020 IEEE

Performance Comparison of BAG and Custom Generated Analog Layout for Single-Tail Dynamic Comparator

Mudasir Bashir, Fatemeh Abbassi, Mirjana Videnovic Misic, Johannes Sturm, Hueber Gernot

Silicon Austria Labs GmbH, Austria

Email: $\{mudasir.bashir, fatemeh.abbassi, mirjana.videnovic - misic, j.sturm, gernot.hueber\}$ @silicon-austria.com

Abstract—The advancements in analog electronic design automation (EDA) tools for layout design are still lagging behind than its counterpart digital EDA tools. Due to which, the analog layout design is still done manually therefore time-consuming and difficult to clear the physical verification checks. Over the recent years, many tools are introduced for the automatic layout generation of analog circuits. This paper presents a study of one such popular layout generator for analog layout design referred as Berkeley Analog Generator (BAG). A step by step workflow of the layout generation in BAG is presented and the performance comparison of the automatic layout generated using BAG and full-custom layout for analog circuits is demonstrated. As a test-case, a typical dynamic comparator is used for comparison. The layout of dynamic comparator is done for the same aspect ratios of transistors and specifications in both environments. The performance validation of layouts is done after parasitic extraction in Cadence Spectre environment using TSMC 28 nm standard CMOS process. A clock signal of 50 MHz for the comparator is considered at a supply voltage of 1.0 V.

Index Terms—Analog integrated circuits, Analog layout, Analog layout automation, Analog layout migration, Analog layout synthesis, Post-layout performance, Technology independence.

I. INTRODUCTION

Analog and mixed-signal (AMS) integrated circuits (ICs) are vital aspect of the modern system-on-chips (SoCs). In the most AMS SoCs, the area occupied by the digital circuits is larger than the analog blocks area, with the effort to implement the latter significantly higher. The automation of digital layout IC designs has become fully matured, technology independent and therefore reusable, while the automation of analog layout IC design is still an open topic due to limited automated synthesis and low IP reuse. Following the CMOS technology downscaling towards multi-patterned technologies, the number of (layout) design rules has increased dramatically (exponentially) making DRC errors debugging time consuming. A way to face these challenges is to pursue reusability and automation of the analog IC layout generation where solving DRC errors and reaching targeted post-layout circuit performance in a swift programmable fashion [1].

Over the last few decades, many attempts have been made to automate analog layout design. None of them were widely accepted as they were not user-friendly (e.g. significant time needed for tool set-up and tool knowledge building). On the other hand, the conventional automatic layout generator tools depend on many constraints, such as device placement,

routing, floor planning thus making the layout optimization complex and time-consuming [2]. The analog layout automation tools that have been introduced over time generate the layout either using top-down based [3, 4], template based [5, 6] bottom-up based [7] optimization methodologies and some use generator-based approaches [8, 9]. In the later approach, a generator is defined for each block or sub-block of the circuit as a piece of code which runs the commands chronologically. These generators work with Matlab, Perl, Python, SKILL or Tcl platform within an environment framework that provides a programming interface between the IC design tools and the process technology kit (PDK). Depending on the generator code a schematic, layout or testbenches for validation can be generated. The generator-based approach has a faster run-time but are less flexible than the optimization approaches. Also, the former approach is knowledge-based and requires in-depth layout and technology knowledge whereas the later methodology can deal with the constraints and multiple dependencies. A combination of both approaches could result in a very efficient automation tool for analog layout design [10].

The paper is organised as follows; Section II introduces the Berkeley Analog Generator (BAG), its workflow and analogies with the manual analog design flow. Section III gives a short review of conventional single tail dynamic comparator. Section IV presents the automated analog layout and manually drawn layout and their post-layout performance comparisons. Section V concludes the paper.

II. BERKELEY ANALOG GENERATOR

BAG is a generator-based tool that develops analog-mixed signal circuits which can be ported from one technology node to another without much effort. The design flow includes parametric design flow which generates a schematic, layout as well as testbenches for verification of the circuits using a set of input specifications [9]. In the typical design flow, a lot of time is spent for choosing the sizing of transistors, plotting the results and then the iterations in the device sizing until the final speculations are achieved. BAG follows the same procedure except here everything is done by a code-based environment hence making the exploration easier and faster. In this paper we will focus only on the layout generation feature of BAG. The same generators can be used across technologies like ST 28 nm FD-SOI, ST 22 nm FDX, TSMC 28 nm, TSMC 16 nm,

978-1-7281-8494-4/20 $31.00 © 2020 IEEE

TSMC 7 nm and Global Foundries 45 nm RF SOI processes without any modifications [9].

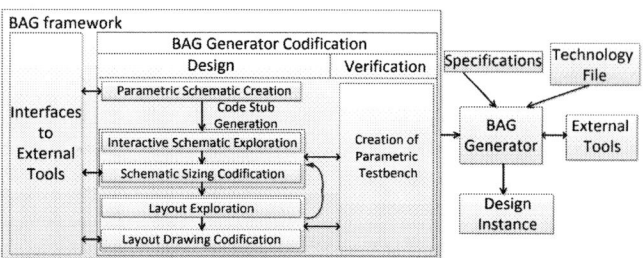

Fig. 1: BAG desigflow [9].

Fig. 1, presents the BAG design flow. It has three different domains, i.e., schematic generation, layout generation and testbench generation. BAG allows to use them separately or together with each other. The schematic generation is custom way of defining the circuit details however here only the connections between the devices is defined and not their geometries or technology nodes. The intention is to reuse the same code for other technologies. In the next phase, test benches are generated. A parametrized file is defined where the simulation setup, parameters and initial conditions are defined. The final phase is to generate the layout. Here, the same procedure as exploration for device sizing at schematic-level is used. Fig. 2 shows the workflow of layout generator in BAG. Since the layout is scripted, it can be modified easily to achieve the target specifications as well as to clear the physical verification checks (Design Rule Checking (DRC) and Layout Versus Schematic (LVS)). After completing the generator, the device sizing, technology parameters and testbench details can be given using a specific input file to get the final design of analog IC.

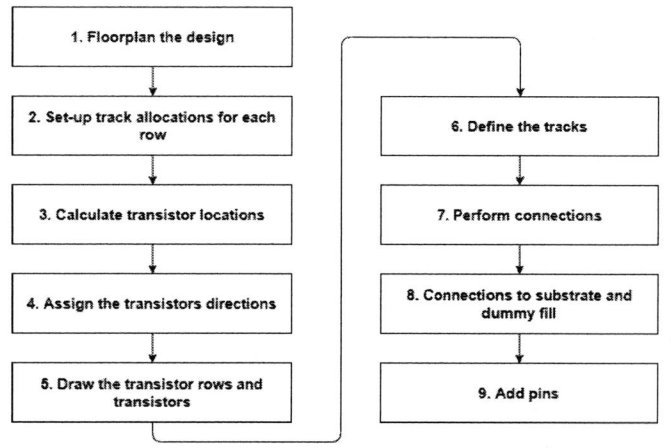

Fig. 2: BAG layout workflow.

III. CONVENTIONAL DYNAMIC COMPARATOR

Fig. 3 shows the schematic of a typical dynamic comparator [11]. They are mostly used due to their advantages such as no static power consumption, high speed, higher input

impedances and rail-to-rail output swing. The operation of the dynamic comparator is explained in Table I. The speed of the comparator can be increased by increasing the aspect ratio of M_{tail}, but then the full amplification can not be achieved since the input transistors (M_1 and M_2) operate in active region for less time. The circuit has better noise performances compared to other topologies and is less sensitive to mismatches, however is not efficient for sub 1-V power supply applications because of transistor stacking issue.

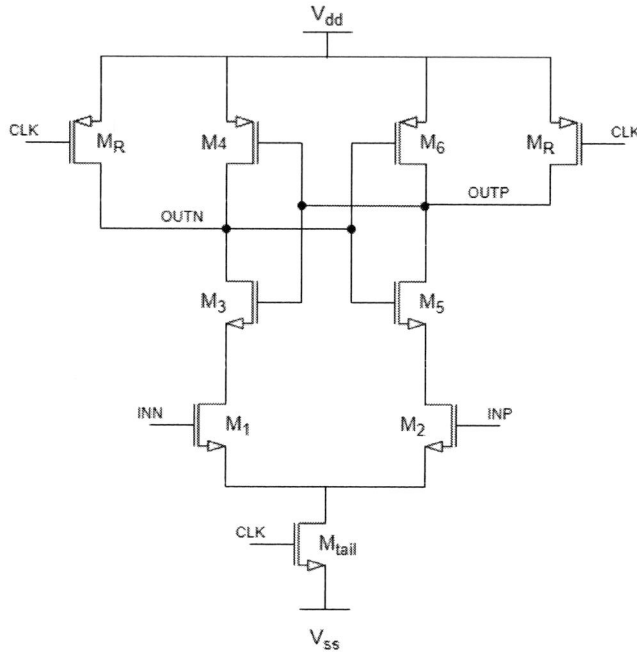

Fig. 3: Conventional Dynamic Comparator [11].

TABLE I: Operation of dynamic comparator

Phase	CLK signal	Operation	Result
Reset phase	$CLK = 0$	M_{tail}= OFF and M_R = ON	Output nodes ($OUTP$ and $OUTN$) are precharged to V_{DD}
Comparison phase	$CLK = V_{DD}$	M_{tail} = ON and M_R = OFF	Output nodes discharge via M_{tail}. For $INN>INP$: $OUTP = V_{DD}$; $OUTN =$ 0 V. For $INN<INP$: $OUTP =$ 0 V; $OUTN = V_{DD}$

IV. RESULTS AND DISCUSSIONS

The layout of the comparator is drawn using the BAG as well as manually. For the sake of demonstration, the comparator is designed in TSMC 28 nm CMOS process at a power supply (V_{DD}) of 1 V. The BAG workflow explained in Fig. 2 is used to generate the layout. Five rows are defined with M_{tail} at the bottom and M_R at the top. The placement of transistors is done in such a way so that better symmetry and matching is achieved in layout design. The same procedure is followed for the manual design except the BAG generates lot of dummy cells to fill the empty spaces within the rectangular box. From the sizing point of view, the width of overall layout design in BAG is decided based on the largest transistor size and the height depends on the number of rows used for device placements. Fig. 4 and Fig. 5 show the layout generated using BAG and the one drawn manually, respectively. The estimated area consumed by BAG generated layout and the manual layout design are 58.4 μm^2 and 36.8 μm^2, respectively (without pad rings).

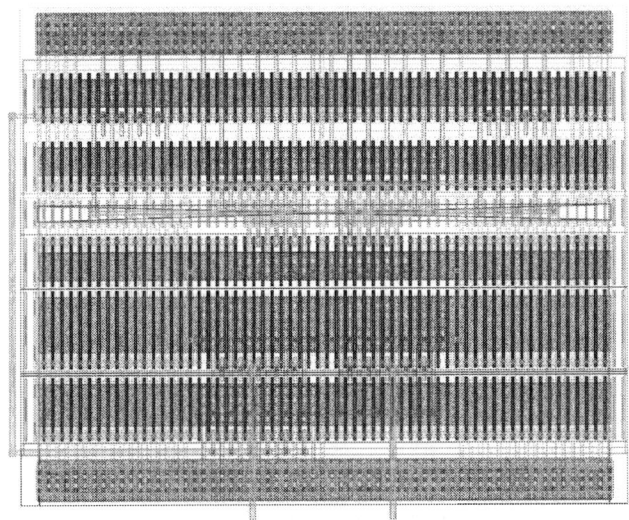

Fig. 4: Layout generated using BAG.

One of the important performance parameter of comparator design is the delay. Fig. 6, Fig. 7 and Fig. 8 show the post-layout results of delay versus common mode input voltage (Vcm), differential input voltage ($Vind$) and power supply (V_{DD}), respectively, for both layout designs. The manual and BAG layout designs result in a delay of 44.77 ps and 43 ps at $Vind$ = 10 mV and V_{DD} = 1 V, respectively. At $Vind$ = 30 mV, the corresponding delays are 37.34 ps and 38.4 ps, respectively. Fig. 9 and Fig. 10 demonstrate the variation of energy per conversion of both these layouts across different $Vind$ and V_{DD} values. A comparison of post-layout simulation results from these layouts at typical-typical corner and (V_{DD}) of 1.0 V is summarised in Table II. The figure-of-merit (FoM) is calculated using Eq. 1, where P_d is power consumption, n is the resolution and f_s is the sampling frequency.

$$FoM = \frac{P_d}{2^n * f_s} \qquad (1)$$

Due to the parallel routing and stacking of higher metal layers in routing, the BAG generated layout has less RC parasitic values on the signal nets at the output nodes of comparator. Therefore, the BAG generated layout results in better delay performance compared to the manual design. The

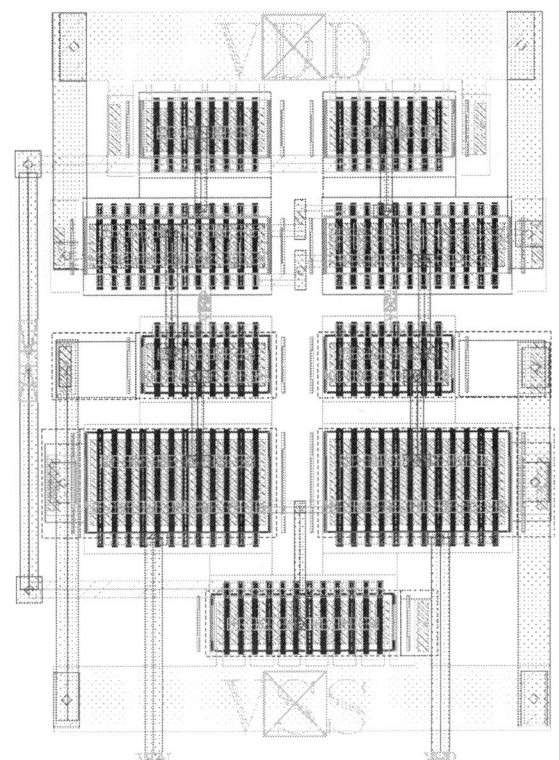

Fig. 5: Layout drawn manually.

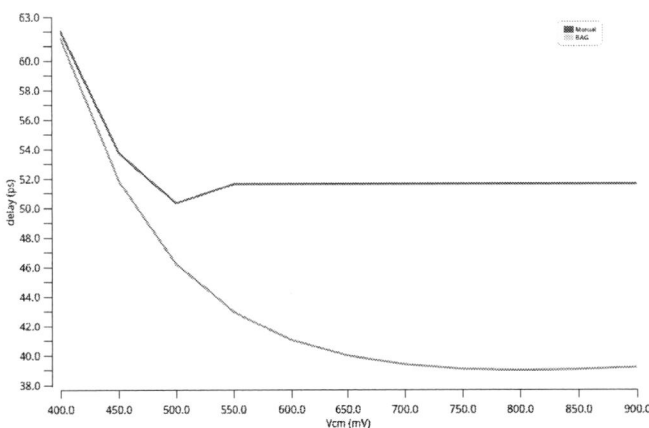

Fig. 6: Delay versus input common mode input voltages.

BAG also generated redundant vias for each layer to avoid any reliability concerns. The BAG generated layout makes it easier to handle the physical verification checks once the environment is set. It takes lot of effort to clear the DRC/LVS errors using manual technique especially for technology nodes ≤ 28 nm. BAG layout generators of a circuit can be used instantly once the environment of a technology node is set up, therefore saving lot of time considering the time taken by analog layout design. Since the BAG generated layout has a very symmetrical structure and contains lot of dummies so at top level it compensates the extra area which is saved in manual layout design at core level. The area of BAG generated

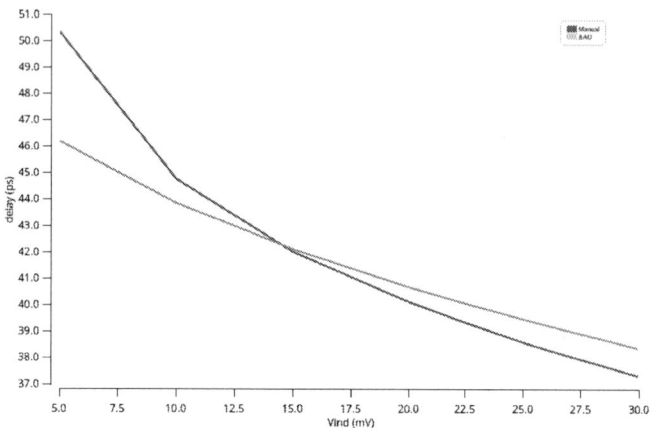

Fig. 7: Delay versus differential input voltages.

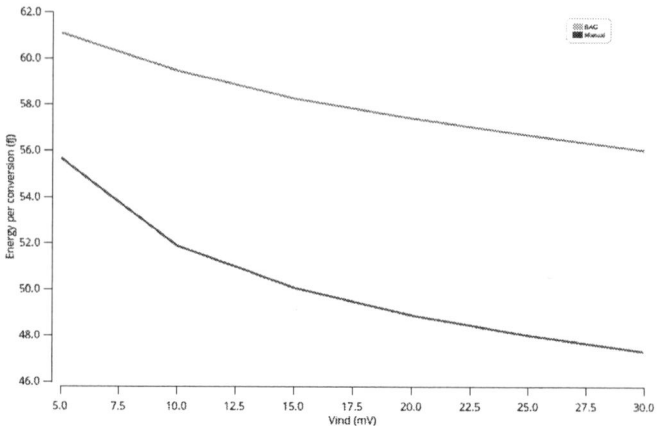

Fig. 9: Energy per conversion versus differential input voltages

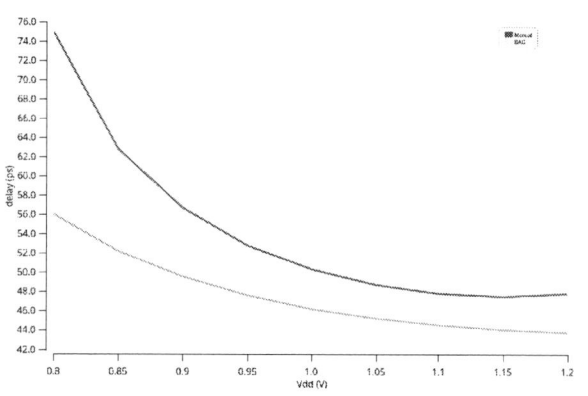

Fig. 8: Delay versus power supply.

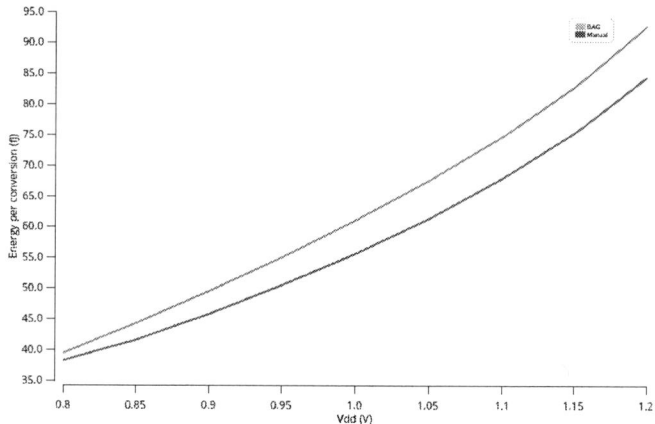

Fig. 10: Energy per conversion versus power supply

layout can be reduced by decreasing the number of rows or changing the positioning of MOS devices. By defining the number of rows as 4 instead of 5 in BAG, the area occupied by the same design is reduced by 30%. As there are no dummies attached with the output net of comparator, the delay performance is improved as well as the power consumption. BAG gives the designers freedom to route and place the devices in layout to achieve better matching, therefore there is more scope to decrease the area and achieve the target specifications. The number of dummies can also be controlled by defining a dummy parameter in the input sizing file.

TABLE II: Post-layout result comparison of BAG and manual layout designs

Parameters	BAG layout	Manual layout	Units
Supply voltage	1	1	V
Technology	tsmc 28	tsmc 28	nm
Sampling frequency	50	50	MHz
Delay	46.21	50.34	ps
Estimated area	58.4	36.8	μm^2
Energy per conversion	61.1	55.68	fJ
FoM	0.94	0.89	fJ/conv

V. CONCLUSION

This paper presents a performance comparison between the automated generated layout using BAG and manual layout. A conventional dynamic comparator is used as an example in tsmc 28 nm CMOS standard process for the performance validations of the efficiency of both these methodologies. The BAG generated layout results similar performances compared to the manual layout design. Although the area consumed by BAG methodology is more than manual design its resuability across different processes makes it more beneficial. The BAG methodology can be very effective during the optimization at top-level floor planning as it takes very less time compared to the manual analog layout optimization.

ACKNOWLEDGEMENTS

The presented research work was done in "Front-End Integrated Circuits and Systems (FEIC), RF systems" research group at Silicon-Austria Labs (SAL) as a part of a cooperative projects with Infineon Austria GmbH.

REFERENCES

[1] H. E. Graeb, *Analog Layout Synthesis: A Survey of Topological Approaches*. New York, NY, USA: Springer, 2010

[2] M. P. Lin, Y. Chang and C. Hung, "Recent research development and new challenges in analog layout synthesis," 2016 *21st Asia and South Pacific Design Automation Conference (ASP-DAC)*, Macau, 2016, pp. 617-622.

[3] R. Martins, N. Lourenço and N. Horta, "LAYGEN II—Automatic Layout Generation of Analog Integrated Circuits," in *IEEE Transactions on Computer-Aided Design of Integrated Circuits and Systems*, vol. 32, no. 11, pp. 1641-1654, Nov. 2013.

[4] H. Habal and H. Graeb, "Constraint-Based Layout-Driven Sizing of Analog Circuits," in *IEEE Transactions on Computer-Aided Design of Integrated Circuits and Systems*, vol. 30, no. 8, pp. 1089-1102, Aug. 2011

[5] R. Castro-Lopez, O. Guerra, E. Roca and F. V. Fernandez, "An Integrated Layout-Synthesis Approach for Analog ICs," in *IEEE Transactions on Computer-Aided Design of Integrated Circuits and Systems*, vol. 27, no. 7, pp. 1179-1189, July 2008.

[6] A. Unutulmaz, G. Dündar, F. Fernández, "Template coding with LDS and applications of LDS in EDA, " *Analog Integr Circ Sig Process*, vol. 78, no. 1, pp. 137-151, 2014.

[7] A. Graupner, R. Jancke and R. Wittmann, "Generator based approach for analog circuit and layout design and optimization," 2011 *Design, Automation & Test (DATE)* in Europe, Grenoble, 2011, pp. 1-6.

[8] B. Prautsch, U. Eichler, T. Reich and J. Lienig, "MESH: Explicit and flexible generation of analog arrays," 2017 *14th International Conference on Synthesis, Modeling, Analysis and Simulation Methods and Applications to Circuit Design (SMACD)*, Giardini Naxos, 2017, pp. 1-4.

[9] E. Chang et al., "BAG2: A process-portable framework for generator-based AMS circuit design," *2018 IEEE Custom Integrated Circuits Conference (CICC), San Diego, CA*, 2018, pp. 1-8.

[10] A. Krinke et al., "From Constraints to Tape-Out: Towards a Continuous AMS Design Flow," 2019 *IEEE 22nd International Symposium on Design and Diagnostics of Electronic Circuits & Systems (DDECS)*, Cluj-Napoca, Romania, 2019, pp. 1-10.

[11] B. Goll and H. Zimmermann, "A Comparator With Reduced Delay Time in 65-nm CMOS for Supply Voltages Down to 0.65 V," in *IEEE Transactions on Circuits and Systems II: Express Briefs*, vol. 56, no. 11, pp. 810-814, Nov. 2009.

Recent Developments in Bandgap References for Nanometer CMOS Technologies

Matthias Eberlein
Institute for Integrated Circuits (IIC)
Johannes Kepler University
Linz 4040, Austria
matthias.eberlein@jku.at

Harald Pretl
Institute for Integrated Circuits (IIC)
Johannes Kepler University
Linz 4040, Austria
herald.pretl@jku.at

Abstract—**This paper provides a short overview about evolving sub-bandgap references, and the related challenges in modern FinFET nodes. Specifically, we present a simple yet innovative concept to generate PTAT and CTAT voltages through "capacitive bias": In replacement of the classical BJT, the active bulk diode is utilized and forward biased by a charge-pump. Two capacitors are discharged across the diode for different time periods, which precisely defines the respective current densities. The sampled diode voltages are then combined by charge sharing or addition, to provide a robust reference. A reverse bandgap reference using this architecture features an untrimmed accuracy of ± 0.73% (3σ), consuming only 21nA in a 16nm FinFET process. The simple structures feature intrinsic supply rejection down to 0.85V and a digital-alike operation.**

Keywords— **Reverse bandgap, Voltage reference, FinFET, bulk diode, N-well junction, switched-capacitor**

I. Introduction

Precise voltage references are a critical backbone of nearly any SoC and will remain indispensable in future technology nodes. They relate closely to thermal sensors, since both IPs rely traditionally on bipolar devices. Other techniques using DTMOS or MOS-only devices are not considered here: Their behavior relies more on process properties than physics, and therefore will always need trimming. A common concept is to generate a signal with positive (PTAT), and one with negative temperature coefficient (CTAT), extracted from the different bias of pn-junctions (Fig. 1). Since this current ratio is typically limited to values of $N = 8 \ldots 20$, the resulting PTAT voltage is small and sensitive to various nonidealities like offset. Classic architectures further include building blocks like amplifiers or current mirrors, which have additional matching requirements. However, analog transistor parameters suffer from the dramatic process scaling, while the IP should fit seamless into a digital system environment.

Fig. 2. Current-mode bandgap circuit proposed by Banba et al. [1].

A. Classic sub-bandgap architectures

Reference circuits do not scale well with latest process nodes, where the limited supply voltage of Vdd ≤ 1.1V is a major challenge. Since the native silicon bandgap is already larger at ~1.25V, new ideas were developed to realize a sub-bandgap operation. By far the most popular reference for low supplies is the "Banba Bandgap" [1], which is based on the current-mode principle (Fig. 2). Due to its simplicity, it is the preferred choice also for recent FinFET nodes, and various publications realize actually adaptions of the same structure [2, 3]. It can operate down to VDD ~ 0.9V and generate flexible output voltages. Major limitations are large resistor values, and mismatch errors from the amplifier and current mirror. It is sensitive to output load (leakage) and supply noise, since the reference is established outside the feedback loop.

Another simple solution was presented by Neuteboom et al. [4], which features also an open-loop configuration (Fig. 3). The current from a standard PTAT source splits into a PNP with series-R, and a parallel resistor. In that way the current through the nonlinear branch features a larger temperature

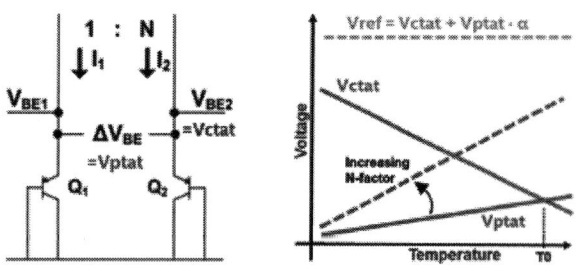

Fig. 1. Vptat and Vctat generation by ratioed biasing (left) and qualitative behaviour to establish a reference (right).

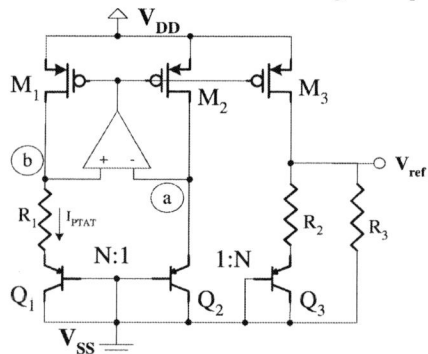

Fig. 3. A sub-bandgap architecture with current division [4].

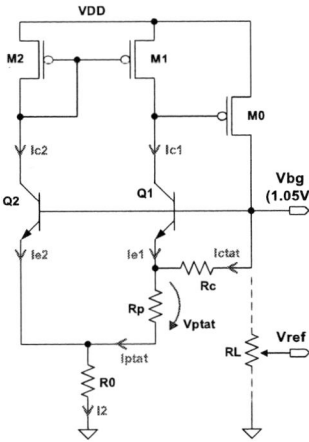

Fig. 4. Closed-loop sub-bandgap reference with NPN devices [6].

coefficient and can achieve reference levels below the native bandgap voltage (Vout ~ 0.8V … 1.1V). A similar approach was presented in [5], using the reverse-short-channel effect in some CMOS technologies to generate a "strong PTAT" signal. However, this solution is expected to be sensitive towards process parameters and likewise requires correction for mismatch errors.

More accurate references are possible if the NPN bipolar device is utilized in an amplifier configuration. The modified Brokaw structure in [6] combines a closed-loop architecture with a boosted PTAT current (Fig. 4). In consequence it achieves good PSRR and precision without trimming in FinFET technology, at low supply headroom. But this requires a triple-well CMOS process, which is not always available.

A separate class of circuits are based on switched-capacitor techniques, still joined with the classic bias of BJT devices. They can compensate offset errors or even do without a differential amplifier, and thereby feature extremely low power consumption [7, 8]. Some more reference structures are summarized in a previous survey [9] and relate mainly to older process nodes.

B. Device constraints in latest technologies

In FinFET nodes a major challenge arises with the quality of bipolar transistors: Traditionally the parasitic PNP is utilized, built from the drain/source, bulk and substrate regions of a standard PMOS device. However, the critical base-emitter junction, unless moved by specific implants, is squeezed inside the fin (Fig. 5). It suffers from various non-uniformity issues, while the base-width Wb is large and fluctuating. This results in very low current gain (<<1), large spread and overall poor I/V characteristic [10].

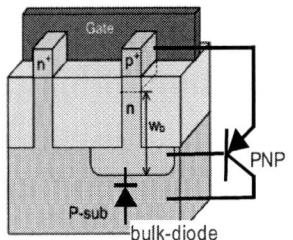

Fig. 5. Parasitic bipolar junctions in a FinFET section with N-well.

The following explanations target especially this critical situation in modern process nodes. We present a solution to replace the classic bipolar transistor with a more predictable and scaling-friendly diode device. In contrast to the static bias of pn-junctions

in prior art, a time-controlled scheme is developed to generate the required bandgap signals with precise temperature coefficients.

II. BASIC PRINCIPLES

A. Bulk diode properties

Recently [11] we introduced the profitable usage of Nwell/Psub diode in forward-bias: Because it is a 2-terminal device, the modeling of electrical characteristics is already simpler and with less parameters. In contrast to parasitic BJTs, this is a buried junction with physical behavior defined from substrate (epi) doping (Fig. 6). Therefore, it is expected to have superior characteristic over conventional drain/source junctions and won't be affected from defects at the surface or fin engineering. Our device analyses in 16nm FinFET demonstrated a near-ideal V/I relationship over 6 decades, down to 10pA. Specifically, the diode voltage spread was observed to be ~ 5 times better than that of the available PNP (Fig. 7).

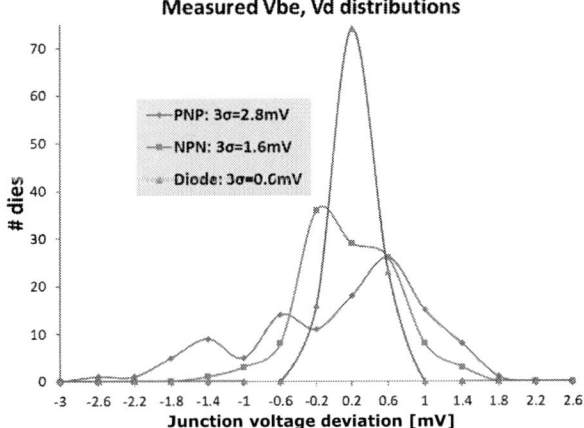

Fig. 7. Measured variability of junction voltages in a FinFET process.

Indeed, there are concerns from classic device physics regarding the ideality factor (η), which is affected from recombination of minority carriers [12]. However, it seems that this effect is getting less relevant in modern technologies with rather shallow wells, but in contrast the η-spread of the PNP transistor becomes problematic. Another distinct advantage of the bulk diode is the availability with virtually any process, while the junction can be sized flexible for good matching. Hence this device is well suited for untrimmed temperature sensors and is highly compatible to future FinFET scaling.

B. Time-controlled biasing of pn-junctions

While traditional circuits bias the bipolar device through a fixed current, the core idea is the utilization of the cap-voltage decay characteristic: A capacitor is initially charged to some voltage > 0.8V (built-in junction potential), and then discharges through the diode. The differential equation for the resulting voltage over time [13] resolves to a simplified formula, after a short initial period (~ 30ns):

$$V_D(t) \simeq \eta \cdot V_T \cdot \ln\left(\frac{C \cdot \eta \cdot V_T}{I_s \cdot t}\right), \qquad (1)$$

where η is the diode ideality factor, VT is the thermal voltage and Is equals the diode saturation current. The

978-1-7281-8494-4/20 $31.00 © 2020 IEEE

minimum (initial) time for discharge depends on the required precision, I_S and serial resistance, and is extracted analytically.

Most important and confirmed by simulation: For several decades of time (t) the diode voltage follows a strict logarithmic function, with nearly no dependence on initial bias (Fig. 8). Consequently, the diode current $I_D(t)$ is also well defined and proportional to 1/t. In our examples we use this relation to generate precise current ratios down to very low levels (< 20nA). Hence, any parasitic series resistors have negligible effect and do not appear in (1). In combination with a nearly ideal diode (Sec. 2.1), this permits further extreme bias ratios of N > 10000, resulting in more robust PTAT voltages (Fig. 1).

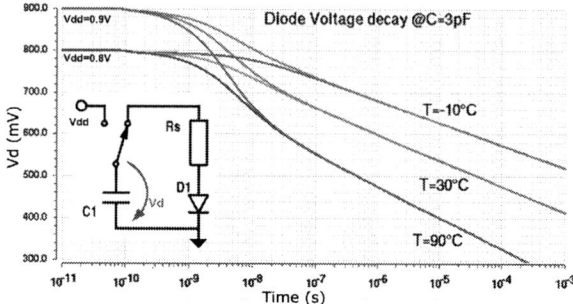

Fig. 8. Simulated diode voltage for different Vdd during capacitor discharge versus time and temperature [11].

C. Forward bias through negative charge-pump

The N-well junction is conventionally utilized for PMOS isolation only and is not a floating diode, because the anode ties to substrate (GND). A straightforward implementation of

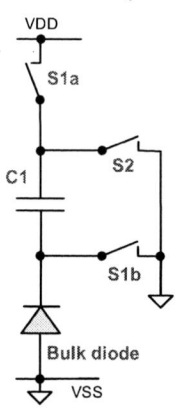

Fig. 9. Switching scheme for diode forward bias.

capacitive (forward) bias to the bulk diode is shown in Fig. 9: Initially the switches S1a/b are closed for pre-charging C1. Thereafter only S2 is closed, which triggers charge-pump operation and pushes the diode cathode to –VDD for immediate discharge of C1. The diode voltage $V_D(t)$, which follows from the specific current (density) $I_D(t)$, is sampled on C1 when S2 opens again, and can be used for further processing. In that way a time-controlled charge-pump is realized, which combines both biasing functions into a single step.

III. APPLICATION TO A SWITCH-CAP REVERSE BANDGAP REFERENCE

Fig. 10 displays the basic circuit, which was introduced originally in [11]. It uses 2 capacitors and a floating diode, while all MOS transistors operate as switches according to the given pulse diagram. The fundamental idea of "capacitive bias" is exploited to define a time-controlled diode current according to (1).

Basically, a reverse bandgap is created from addition of PTAT and CTAT signals, scaled by a factor k:

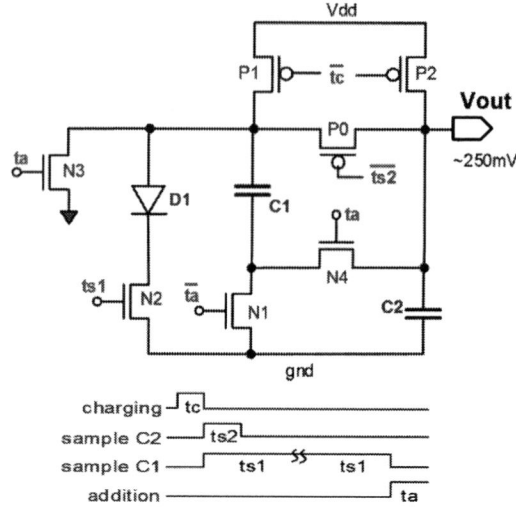

Fig. 10. A time-controlled bandgap example with a floating diode, and associated control pulses [11, 15].

$$Vrbg = Vptat + Vctat = \Delta Vd + Vd/k$$

Applying the capacitive bias principle, with diode voltages sampled at times ts2 and ts1 (with ts1 >> ts2), we can therefore express the reference voltage also from CTAT parts:

$$Vrbg = Vd(ts2) - Vd(ts1) \cdot \left(1 - \frac{1}{k}\right) \quad (2)$$

A. Parallel circuit scheme

The architecture in Fig. 10 is a direct implementation of (2) by switched-capacitor techniques. Specifically, the inherent PTAT voltage is defined precisely by time ratios (1), and may adopt particular large values:

$$Vptat = \eta V_T \cdot ln\left(\frac{C2}{C1} \cdot \frac{ts1}{ts2}\right) \quad (3)$$

The CTAT part is directly related to ln(t) according to (1), but the clock needn't be precise. Such reference could also run from a plain RC-oscillator, since the process spread of the capacitors cancels out if matched.

The switching scheme in Fig. 11 explains the operation of this solution using charge-sharing, as per the schematic (Fig. 10): In the initial phase C1 and C2 are charged to Vdd (or any other voltage larger than the built-in junction potential of the diode). Thereafter C2 (or both capacitors) discharge over the diode for a short time ts2. During phase 2b the voltage Vd(ts2) is sampled at C2, and only C1 continues to discharge for a much larger period ts1. In the last phase the reference voltage is generated, when Q(C2) merges with −Q(C1), which results in scaled CTAT parts according to the capacitor ratios:

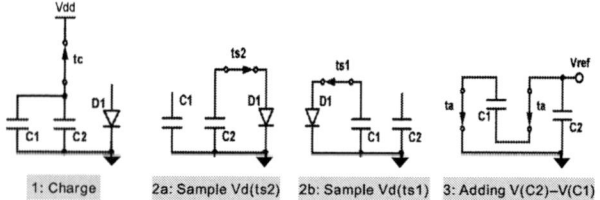

Fig. 11. Capacitor switching phases to generate a bandgap voltage Vref with sampling timings tc, ts1, ts2 and ta [11].

978-1-7281-8494-4/20 $31.00 © 2020 IEEE

$$Vref = \frac{C2}{C1+C2} \cdot \left[Vd(ts2) - \frac{C1}{C2} \cdot Vd(ts1) \right] \quad (4)$$

Clearly, this implements a reverse bandgap as per (2). However, the voltage level is reduced by a damping factor, due to the charge sharing. Repeated operation of phases 1-3 will provide a stable reference by using a sampling capacitor at the output.

B. Serial circuit scheme

Another circuit variant is possible by serial connection and scaling of the capacitor voltages [14]. Fig. 12 shows the transistor level realization using a single bulk diode. Since a floating diode is not available in standard CMOS, it combines time-controlled bias with an instant charge-pump operation, required to generate forward-bias for D1.

Operation is as follows: During initialization (tc) the capacitors C1 and C2 are charged, while C3 is discharged (switches M1, M2, M5, M6 closed). In the second phase node B is pushed to −VDD by closing M3 and M4 only, and both capacitors start discharging through the diode. The respective voltages are sampled on C2 and C1, during times ts2 (M4 opens) and ts1. Voltage subtraction happens in phase 3 (ta), where node C stays at 0V (M3 closed) and −V(C1) is at once scaled down by charge sharing with C3 (M1 closed). The resulting output voltage is an equivalent implementation of (2) and calculates to:

$$Vref = Vd(ts2) - \frac{C1}{C1+C3} \cdot Vd(ts1)$$

$$= \eta V_T \left(ln \frac{(C1+C2) \cdot \eta V_T}{Is \cdot ts2} - \frac{C1}{C1+C3} \cdot ln \frac{C1 \cdot \eta V_T}{Is \cdot ts1} \right) \quad (5)$$

This scheme avoids the damping factor related to charge-sharing and yields the highest reference level, around 380mV in our example. Here we used quite large capacitors (C1 = 1.7pF, C2 = 4pF, C3 = 1pF), and sampling times of ts1 = 100µs and ts2 = 50ns. This produces large PTAT levels with an effective current ratio of N ~ 6700. The clock frequency (20MHz) is not critical, since deviations affect only the CTAT portion slightly. To prevent latchup, M1 and M2 must be PMOS devices and are driven by control signals LS(t) with an offset of −VDD. The level-shifting may be realized by simple circuitry, as shown in Fig. 12.

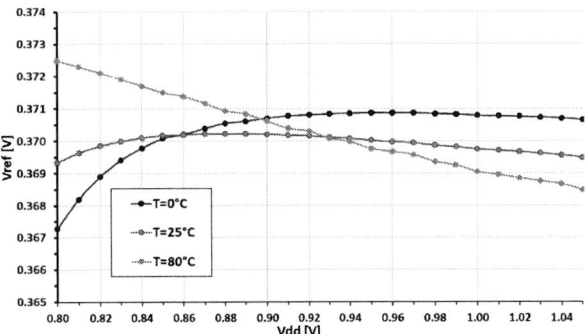

Fig. 13. Measured Vref vs. VDD across temperatures for a typical sample.

C. Silicon realisation and constraints

Successful prototypes have been implemented in 16nm FinFET [11, 14], and results from the serial circuit version are presented here: Fig. 13 depicts the measured line regulation for various temperatures. It confirms the intrinsic robustness against supply variation and functionality down to ~0.8V, as predicted by the capacitive bias concept (Sec. 2). From lab measurements the bandgap voltage showed a total spread of < 0.8% across temperature (Fig. 14), without trimming. Here, the clock frequency was adjusted to cope with (diode) model uncertainties (batch calibration). Hence, the results don't exhibit the typical curvature, but reflect some secondary effects. Eventually, the circuit achieves a small silicon footprint of 1680µm², and a low power consumption of ~ 18nW (including switch drivers). Table 1 summarizes the performance parameters and compares with related work.

This "serial version" is simpler than the parallel scheme, with less constraints on the switch design. Due to the large capacitors used, the impact of mismatch is negligible, while the pulse widths (ts1, ts2) are precise by design. The bandgap accuracy is further enhanced from the higher output level. However, this circuit is sensitive to leakage at node A (Fig. 12), because the effective capacitance driving Vref is small from C2 serial connection. Also, parasitic capacitors can have a relevant scaling impact and should be handled by careful layout extraction. In general, the architecture is robust against clock feedthrough and supply variation. Concerns about substrate noise are tackled effectively by a diode guard ring.

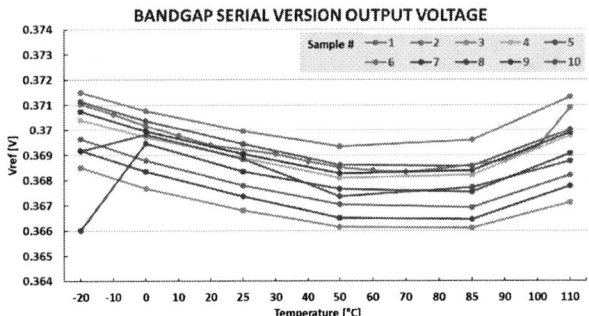

Fig. 13. Measured Vref vs. VDD across temperatures for a typical sample.

IV. CONCLUSION

For decades of IC design, the bandgap reference adapted successfully to the evolving process nodes. Features like sub-1V operation or curvature correction were implemented with rather small architectural changes. But with entering the

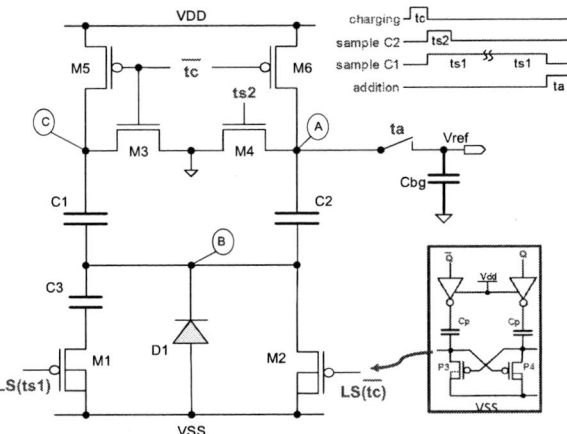

Fig. 12. Serial version of the SC-Reverse Bandgap with a single bulk-diode. IO-devices are indicated with a thick gate, to reduce leakage.

FinFET age, the parasitic BJTs are no more "good enough" to yield acceptable performance. Our proposal to utilize bulk diodes represents a new paradigm, as they do not suffer from technology scaling and future fin engineering. Initial measurements have shown, that they perform better than BJTs in various aspects. Combined with capacitive bias, this enables very compact, low power and process-insensitive circuits, which may be calibrated conveniently by pulse timings.

A major advantage is simplicity, which receives increasing value recently [15]: This digital-like concept dispenses with classic structures of bias, current mirror or feedback loop. It neither relies on analog transistor performance like matching, or on specific process options such as resistors or MIM-caps. The core circuitry features intrinsic supply rejection and can easily run directly from sub-1V sources or a small voltage doubler. Based on this idea a reverse bandgap reference was realized on $1680\mu m^2$ silicon area, which can operate at low supply of VDD $\geq 0.85V$. It achieves a 3 σ -error of $\leq 0.82\%$ without calibration and consumes only 21nA of current, including the switch drivers. The same principle may be applied for other circuits like temperature sensors, which rely heavily on bipolar devices [16].

REFERENCES

[1] H. Banba *et al.*, "A CMOS bandgap reference circuit with sub-1-V operation," in *IEEE Journal of Solid-State Circuits*, vol. 34, no. 5, pp. 670-674, May 1999.

[2] Y.-W. Chen, J.-J. Horng, C.-H. Chang, A. Kundu, Y.-C. Peng and M. Chen, "A 0.7V, 2.35% 3σ-Accuracy Bandgap Reference in 12nm CMOS," in *IEEE ISSCC Dig. Tech. Papers*, USA, 2019, pp. 306-307.

[3] C. H. Chang, J. J. Horng, A. Kundu, C. C. Chang and Y. C. Peng, "An ultra-compact, untrimmed CMOS bandgap reference with 3σ inaccuracy of +0.64% in 16nm FinFET," in *Proc. IEEE A-SSCC*, Kaohsiung, Taiwan, 2014, pp. 165-168.

[4] H. Neuteboom et al, "A DSP- based hearing instrument IC," in *IEEE Journal of Solid-State Circuits*, vol. 32, pp. 1790–1806.

[5] A. Annema and G. Goksun, "A 0.0025mm² bandgap voltage reference for 1.1V supply in standard 0.16μm CMOS, *" 2012 IEEE International Solid-State Circuits Conference (ISSCC)*, San Francisco, CA, 2012, pp. 364-366.

[6] M. Eberlein and H. Pretl, "A Low-Noise Sub-Bandgap Reference with a ±0.64% Untrimmed Precision in 16nm FinFET," *in Proc. A-SSCC*, Macao, 2019, pp. 113-116.

[7] V. Ivanov, R. Brederlow and J. Gerber, "An Ultra Low Power Bandgap Operational at Supply From 0.75 V," *IEEE J. Solid-State Circuits*, vol. 47, no. 7, pp. 1515-1523, July 2012.

[8] A. Shrivastava, K. Craig, N. E. Roberts, D. D. Wentzloff and B. H. Calhoun, "A 32nW bandgap reference voltage operational from 0.5V supply for ultra-low power systems," in *IEEE ISSCC Dig. Tech. Papers*, San Francisco, CA, USA, 2015, pp. 1-3.

[9] C. Fayomi, G. Wirth, H. Achigui and A. Matsuzawa, "Sub 1V CMOS bandgap reference design techniques: a survey". *Analog Integrated Circuits and Signal Processing*, vol. 62, pp 141-157, Feb. 2010.

[10] U. Kamath et al., "BJT Device and Circuit Co-Optimization Enabling Bandgap Reference and Temperature Sensing in 7-nm FinFET," in *Proc. ESSDERC*, Dresden, 2018, pp. 86-89.

[11] M. Eberlein, G. Panagopoulos and H. Pretl: „A 40nW, Sub-1V Truly 'Digital' Reverse Bandgap Reference Using Bulk Diodes in 16nm FinFET", in *A-SSCC Tech. papers*, Tainan, Taiwan, pp. 99-102.

[12] M. A. P. Pertijs and J. H. Huijsing, "Precision temperature sensors in CMOS Technology". New York, NY, USA: Springer, 2006, pp. 125–128.

[13] E. Hellen, "Verifying the diode-capacitor circuit voltage decay", American Journal of Physics, vol. 71, August 2003.

[14] M. Eberlein, H. Pretl and Z. Georgiev, "Time-Controlled and FinFET Compatible Sub-Bandgap References Using Bulk-Diodes," in *IEEE TCAS II: Express Briefs*, vol. 66, no. 10, pp. 1608-1612, Oct. 2019.

[15] A. Altvater, "Matthias Eberlein Honored With the IEEE Brokaw Award for Circuit Elegance," in *IEEE Solid-State Circuits Magazine*, vol. 12, no. 2, pp. 80-81, Spring 2020.

[16] M. Eberlein and H. Pretl, "A No-Trim, Scaling-Friendly Thermal Sensor in 16nm FinFET Using Bulk Diodes as Sensing Elements," in *IEEE Solid-State Circuits Letters*, vol. 2, no. 9, pp. 63-66, Sept. 2019.

TABLE I.   PERFORMANCE SUMMARY

Silicon Data	This work (serial concept)	[2] (Chen)	[3] (Chang)	[8] (Ivanov)	[9] (Shrivastava)
Process	16nm FinFET	12nm FinFET	16nm FinFET	130nm	130nm
VDD	0.85V - 1.05V	0.7V	1.8V	0.75V	0.5V
Power	18nW	6.9μW	-	170nW	32nW
Vref	370mV	207mV	615mV	185mV	423mV
Spread (3σ) w/o trim	± 0.73% (min/max)	± 2.35%	± 1.67%	± 0.9%	± 2.0%
PSRR	~ 49 dB (at 25°C)	-	-	~ 53dB	40dB
Area	0.0016 mm²	0.065mm²	0.0023mm²	0.07mm²	0.0264mm²
# Samples	10	35	63	~ 300	6
Type	Reverse BG	Current-mode BG	Current-mode BG	Reverse BG	Fractional SC BG

Comparison of EMI Improved Differential Input Pair Structures Within an Integrated Folded Cascode Operational Transconductance Amplifier

Dominik Zupan, Bernd Deutschmann
Institute of Electronics (IFE)
Graz University of Technology
Graz, Austria
dominik.zupan@tugraz.at, bernd.deutschmann@tugraz.at

Abstract—This paper investigates the electromagnetic interference of integrated folded cascode operational amplifiers. In this context seven electromagnetic interference improved differential input pair structures are compared against each other in terms of their susceptibility, but also in terms of their influence on an initial standard reference circuit design and performance. Special focus is laid on not significantly altering these initial specifications that are defined for an integrated folded cascode amplifier, that is used as a case study. Comparisons are made of the investigated structures in terms of electromagnetic interference rejection ratio, area requirement, power consumption, changes in differential and common mode gain, gain-bandwidth product etc.

Index Terms—electromagnetic interference, operational amplifier, EMI improved, EMI robustness, folded cascode, EMI induced offset, differential input pair

I. INTRODUCTION

Robustness against electromagnetic interference (EMI) is a very critical aspect in modern integrated circuit (IC) designs. Conducted, coupled or radiated disturbances can lead to distortion phenomena, direct current (DC) shift and other effects hindering electronic circuits and devices to function properly. A severe effect in operational amplifier structures is a change in output voltage due to EMI, as such amplifiers are often used as input structures for sensitive analog signals, e.g. sensor interfaces. Such analog signals are often provided by wires in a cable harness or by printed circuit board (PCB) traces, and can therefore be overlaid with interference signals, as PCB traces and cables might act as receiving antennas for EMI.

There are already many different techniques described in the literature, offering concepts to reduce the effects of EMI on operational amplifiers. A recommended sequence on how to approach EMI problems is given in [1]. Therefore known countermeasures can be split into concepts of filtering [2], linearisation [3], [4] and compensation [5], [6], [7]. Most of the EMI countermeasures have the objective of cancelling disturbances coming into an IC via the inputs. These are often very complex to counteract [8, pp. 142-145]. The differential input stage plays a critical role in this, as changes of the operating point (OP) in this first stage are passed on to the next stages and would need to be counteracted subsequently. Making this first stage more EMI robust is thus critical for the

design. Most of the existing literature focuses on comparing a classical amplifier structure with an EMI improved structure of the same amplifier type. There is little information available, however, on the effects of different EMI improved input pairs in relation to each other. Furthermore, information on the influence of these different structures on the specifications of classical amplifier structures is scarcely available in any form.

This paper puts its focus on a comparison of known measures against EMI, implemented in a $0.35\,\mu m$ technology, for the reasons mentioned, and does so not only in terms of effectiveness, but also in terms of area requirements, the need of matching or exact values, power consumption etc. Additionally changes on the initial amplifier specifications are investigated in a case study. This should help designers in their decision-making process when they must make a trade-off between initial specifications and the robustness requirement.

This paper is organised as follows: In section II a classical folded cascode operational amplifier, its specifications and also its susceptibility to EMI is described. In Section III seven different EMI measures on the differential input pair structure are revisited. Section IV deals with the analysis and comparison of these structures in terms of their EMI robustness, but also with the investigation of changes to the nominal specifications defined in Section II.

II. FOLDED CASCODE OPERATIONAL AMPLIFIER AND EMI CONSIDERATIONS

A folded cascode operational amplifier is used as a case study to investigate the effectiveness of different EMI robust differential input pairs. Therefore this section summarises the used circuit and highlights its EMI problems.

The basic reference schematic, called structure *A* further on, is shown in Figure 1, whereas the sizing of transistors M_0 to M_{10} is, in principle, not subject to change. This is done in order to keep the applied methods comparable. However, for some input structures the sizing has to be adapted. The basic folded cascode operational amplifier is designed to have an open loop gain of about $80\,dB$, a gain-bandwidth product (GBWP) of about $10\,MHz$ and a phase margin of about $72°$ at an OP of $V_{inp} = V_{inn} = \frac{V_{DD}-V_{SS}}{2}$. A summary of all specifications is given in Table I, column *A*.

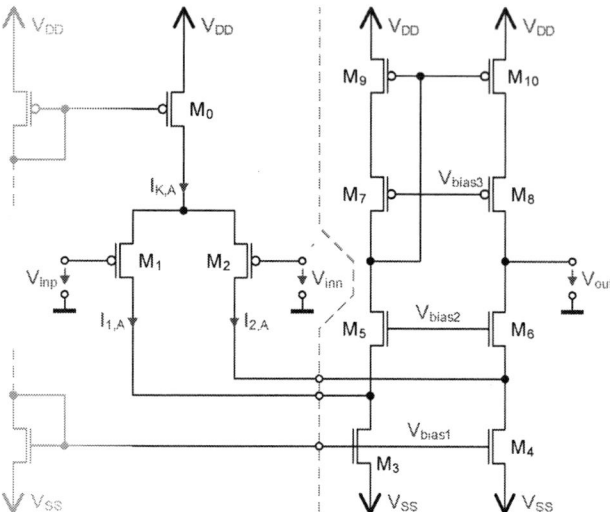

Fig. 1. Structure A: Standard initial folded cascode operational amplifier used as reference. In the course of the paper the input stage will be replaced by EMI improved structures.

The goal is to investigate the change in offset voltage at the output due to EMI disturbances. Two offset types thus need to be distinguished:

- The technological offset, which is introduced by the design and layout/matching of the amplifier and can be seen as intrinsic characteristic. [1]
- The EMI induced offset arises by contrast, when an EMI signal couples into the pins of the amplifier and temporarily alters its DC OP.

Measures for reducing the technological offset need not necessarily have an effect on the EMI induced offset, which was shown in [9], [10]. This is the case due to different origins of offset generation. The technological offset mainly occurs due to non-perfect matching, whereas the EMI induced offset occurs due to non-linearities in the circuit. EMI can couple into such ICs via either input pins, the output pin or also the supply pins. However, the main focus in this case study will lie on the input pins, as these are most complex to counter [8, pp. 144-145], and especially the non-inverting input pin, when operating the amplifier as a voltage follower. This is for example of special interest, when operating the amplifier as sensor interface, where EMI could couple into PCB lines or cable harnesses of attached sensors.

There are different ways an EMI induced offset can orginate due to disturbances [8, pp. 144-145]:

- Slew rate asymmetry, which is not dealt with here, as the amplifier is designed to have equal slew rates (see Table I).
- Strong non-linear distortion on the input stage, such that the transistors are forced into the cut-off or the ohmic region and
- weak non-linear distortion on the input stage. Whereas the latter is the most important EMI effect and is focused

---

[1] The technological offset is not focused on in this paper, as there are already well-defined compensation techniques available, such as chopping or auto-zeroing.

on, as it is already occurring due to small disturbances. Weak non-linear distortion can already be seen in the simplest level 1 transistor models, due to the drain current being quadratically proportional to the overdrive voltage ($V_{GS} - V_{TH}$). Due to this non-linear behaviour the drain currents are distorted. The offset voltage does not change much for in-band signals due to the high gain feedback loop. In the case of out-of-band signals, however, the amplifier can be seen as being operated in open loop, which leads to an EMI offset voltage being induced and therefore to a temporary, disturbance dependent output voltage change.

Different measures to mitigate these mentioned effects due to EMI injection are presented in the following. As already stated a special focus will be placed on adding EMI robustness, without significantly altering the circuit's behaviour and characteristics.

III. EMI Robustness Measures

This section deals with measures and adaptions to the differential input pairs, in order to make the folded cascode amplifier more immune to EMI. In general there are four main concepts for increasing the EMI robustness of these input pairs adapted from [1]: filtering, linearisation, bandwidth and compensation. Each of these are explained in the subsequent sections.

A. Filtering

Filtering represents one of the simplest, but also most effective ways of eliminating unwanted EMI input signals. The influence of high frequency distortions is decreased, by adding filters to both inputs. In this way EMI signals should not reach the input differential pair at all. Such filters can be implemented in two ways:

- Externally: There are several issues that can arise when using external filters on PCB level. One crucial factor is that the two filter structures cannot be precisely matched, which may lead to a technological offset. Furthermore, the increased bill of material (BOM) leads to higher costs in production.
- Internally: Using internal/on-chip filtering reduces the previously stated problems of matching and BOM. However, due to the need of a low cut-off frequency of the filter, the on-chip area requirement is significantly increased.

The filtering input structure (used internally), that is investigated in this paper is subsequently denoted as structure B and depicted in Figure 2. Its effectiveness is discussed in [2]. The filtering (R_B, C_B) is designed to have a cut-off frequency of about $50\,\mathrm{MHz}$. Keeping the same cut-off frequency, the area of the low-pass filter area can be decreased by increasing R_B, while simultaneously decreasing C_B. By contrast, however, the noise introduced by R_B increases with its value. This leads to a trade-off between area and noise. In this case study the noise contribution of the input transistors and the resistance are made to approximately match, which leads to a resistance of about $R_B = 1\,\mathrm{k\Omega}$ and a capacitance of about $C_B = 3.2\,\mathrm{pF}$.

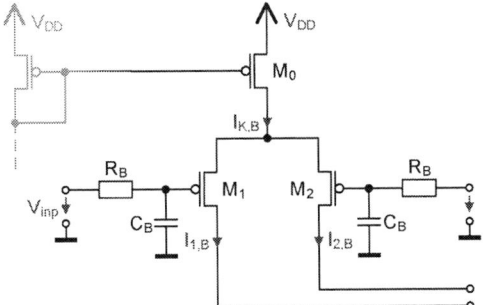

Fig. 2. Structure B: Differential pair with input RC-filter.

B. Linearisation

When filtering is not applicable or practicable, the next measure suggested by [1] is linearisation. In this way the effects of weak non-linear distortions are minimised, making both of the drain currents linearly proportional to the overdrive voltage in a specific input voltage range. Three approaches of linearising the differential input stage are summarised in the following.

1) Source Degeneration: Adding a resistor to the input differential pair source is a relatively simple approach, and it already helps in linearising the input stage [8, p. 154], [3]. By this means, however, the transconductance is also lowered, when increasing R_C and therefore the differential gain A_{DM} also decreases. The structure of the source degenerated differential pair is given in Figure 3 as structure C. When dimensioning the resistor R_C a trade-off between obtaining a higher EMI rejection (larger resistor size) and maintaining a higher differential gain (smaller resistor size) has to be found. In this case study $R_C = 1\,\text{k}\Omega$ is used as trade-off.

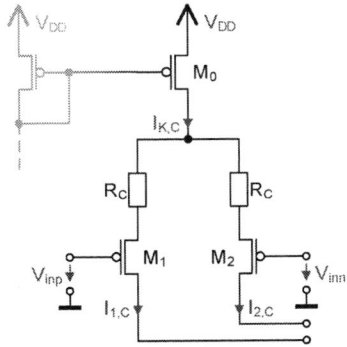

Fig. 3. Structure C: Differential pair with resistive source degeneration.

2) Cross-Coupled Differential Pair: Another approach in linearising the transconductance of the input structure utilises a second, cross-coupled differential pair, as shown in Figure 4, structure D. The third-order harmonic of the unfiltered input signal is cancelled out, when equation 1 is fulfilled [4]. In order to maintain the same differential gain, the transistor dimensioning needs to be adapted (increased width), in such a manner that the equations $i_{K,A} = i_{K,D1} - i_{K,D2}$ and therefore $i_{1,D} = i_{1,A}$ and $i_{2,D} = i_{2,A}$ hold true for the small-signal currents.

$$\frac{I_{K,D1}}{I_{K,D2}} = \left(\frac{\frac{W_{1,D1}}{L_{1,D1}}}{\frac{W_{1,D2}}{L_{1,D2}}} \right)^3 \qquad (1)$$

The currents $I_{K,D1}$ and $I_{K,D2}$ are chosen to have a ratio of about 3. This leads to width over length ratios of about $\frac{W_{1,D1}}{L_{1,D1}} = \frac{W_{2,D1}}{L_{2,D1}} = \frac{200\,\mu\text{m}}{2\,\mu\text{m}}$ and $\frac{W_{1,D2]}}{L_{1,D2}} = \frac{W_{2,D2]}}{L_{2,D2}} = \frac{140\,\mu\text{m}}{2\,\mu\text{m}}$. However for DC the overall current is consisting of $I_{K,D1} + I_{K,D2} > I_{K,A}$. Therefore M_3 and M_4 of the initial reference design have to be extended by the transistors $M_{3,D}$ and $M_{4,D}$ in order to cope with the increased DC currents $I_{1,D}$ and $I_{2,D}$. The main disadvantage with this concept is the need for proper matching of each of the transistors in the two differential pairs, as well as proper matching among the two differently dimensioned differential pairs.

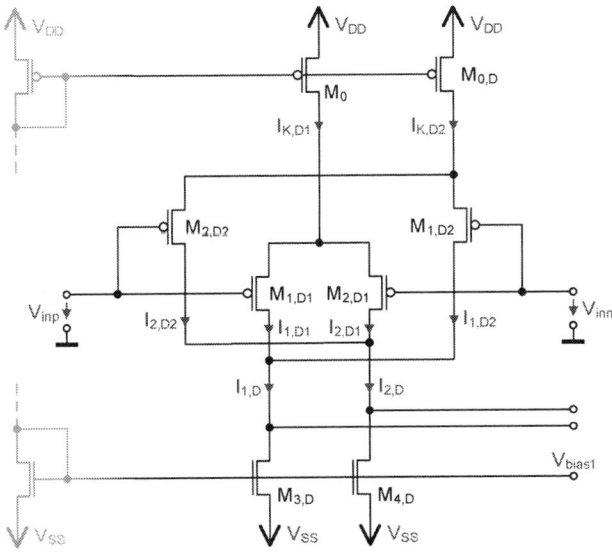

Fig. 4. Structure D: Differential pair with cross-coupled double differential pair.

3) Combination: Source Degeneration and Cross-Coupled Differential Pair: Finally a combination of source degeneration and cross-coupled double differential pairs can be investigated as shown in Figure 5 (structure E) and dealt with in [4]. This leads to a wider input range of linear behaviour. However, the differential gain is again decreased, as the transconductance g_m is decreased too. Again the initial transistors M_3 and M_4 need to be extended by the transistors $M_{3,E}$ and $M_{4,E}$ similarly to structure D. The resistors are sized by $R_D = 500\,\Omega$, as a trade-off between gain and linearisation.

C. Increasing Bandwidth

Increasing the bandwidth/speed of the amplifier helps in keeping the feedback loop stable at higher frequencies. By this means the offset voltage would intrinsically be partly compensated. However, as the goal of this paper is to maintain the given characteristics of structure A the bandwidth is not subject to change.

978-1-7281-8494-4/20 $31.00 © 2020 IEEE

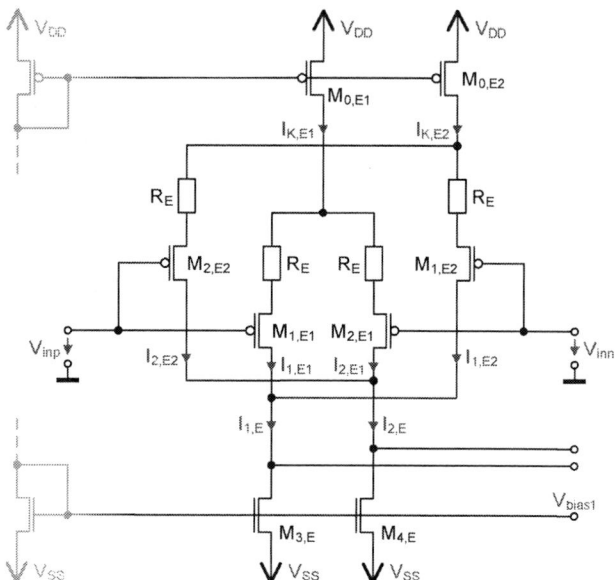

Fig. 5. Structure E: Differential pair with cross-coupled double differential pair.

D. Compensation

The last measure of [1] consists of compensating distortions, that are introduced by EMI, which occur, e.g. if the input filter is not that effective to mitigate an EMI disturbance.

1) Cross-Coupled Double Differential Pair: Again a cross-coupled double differential pair can be used for compensation of EMI effects in the input stage. However, in contrast to structure D no linearisation is happening, but the non-linear distortion is occuring first and is subsequently cancelled out [5]. This is done by an equally-sized cross-coupled differential pair ($M_1 = M_{1,F}$, $M_2 = M_{2,F}$), that is only activated at higher frequencies by using high-pass filters, which are designed to have the same corner frequency (and dimensioning due to the same considerations) as the low-pass filters from section III-A. The idea is, that at frequencies above the corner frequency of the high-pass filter, the current difference introduced by M_1 and M_2 is exactly counteracted by the transistors $M_{1,F2}$ and $M_{2,F2}$, such that no current difference is introduced. By this means no high-frequency EMI signal should pass through to the output, since it is now compensated and counterbalanced. The schematic of this input stage is illustrated in Figure 6 and denoted as structure F. Again $M_{3,F}$ and $M_{4,F}$ are added to cope for the higher DC currents through both branches. The bias voltage V_{bias4} is adapted in such a manner that the transistor $M_{1,F2}$ and $M_{2,F2}$ are operated in saturation.

2) Complementary Double Differential Pair: A complementary double differential pair can be used similarly to structure F. Again the additional differential pair is only active at higher frequencies (by using high-pass filters), aiming for compensation of the non-linearly distorted differential current [6]. In this case again, the differential pairs are not linearised but the current difference $I_{1,G1} - I_{2,G1}$ is counteracted by the current $I_{1,G2} - I_{2,G2}$ at higher frequencies. To account for the higher currents through M_9 and M_{10} the dimensioning

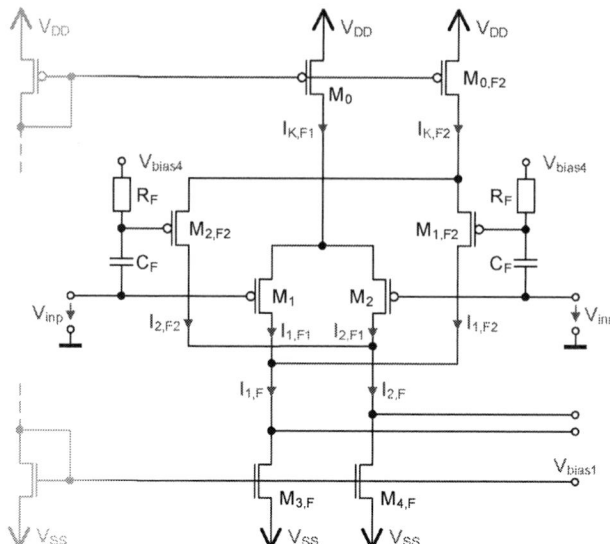

Fig. 6. Structure F: Differential pair with cross-coupled double differential pair for compensating non-linear effects.

has to be adapted by adding the transistors $M_{9,G}$ and $M_{10,G}$ in parallel. This is illustrated in Figure 7 as $M_9 + M_{9,G}$ and $M_{10} + M_{10,G}$, meaning the sum of widths W_9, $W_{9,G}$ respectively W_{10}, $W_{10,G}$.

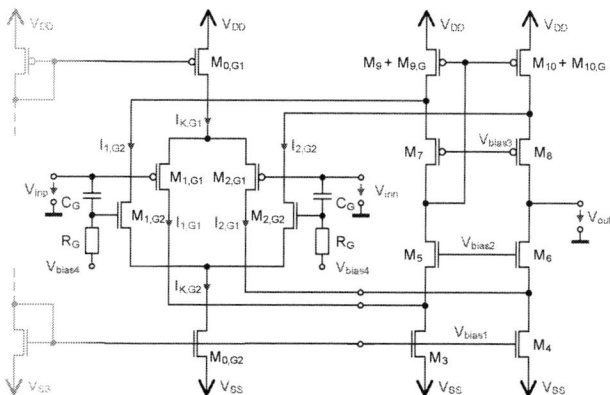

Fig. 7. Structure G: Differential pair with complementary double differential pair for compensating non-linear effects.

3) Source-Buffered Differential Pair: In contrast to the previous structures, source-buffering aims at achieving a different type of compensation. In this case the back-gate effect is utilised, as described in [7]. Along with specific dimensioning of the gate-source voltage by adding C_H the common-mode transfer function of the amplifier is forced to zero. The associated circuit is denoted as structure H and is shown in Figure 8. However, in this input pair the problem of needing exact values (not only matching) arises. Deviations from this ideally calculated capacitance have significant influence on the performance of this structure.

IV. SIMULATIONS AND RESULTS

In this section the performed simulations are explained and the results are analysed, by comparing the structures A to H,

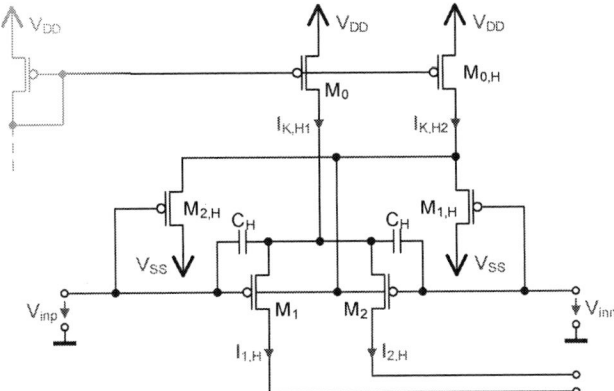

Fig. 8. Structure H: Differential pair with source-buffering.

that were described in the previous section. All characteristics are measured at an voltage of $V_{DD} = -V_{SS} = 1.65\,\text{V}$, at a nominal temperature of $27\,°\text{C}$ and for nominal corners.

On the one hand typical amplifier characteristics such as technological DC offset, gain, GBWP, area, power consumption etc. are compared, in order to ensure, that any changes to the input stages do not significantly alter the amplifier's behaviour. On the other hand the structures are compared concerning their EMI robustness. All of these are summarised in Table I and are described within this section. In addition, the table cells are colorised, in order to represent good/bad effects in terms of deviations from the reference specification.

Evaluating the EMI robustness can be done by calculating the so called electromagnetic interference rejection ratio (EMIRR) [11], which is commonly used, despite its flaws [12], [13]. For comparing the behaviour of the folded cascode circuits in closed loop, the EMIRR of the non-inverting input pin (EMIRR IN+) is measured as described in [11, pp. 4, 5]. For this case the DC input voltage is superimposed by a high frequency signal with defined peak voltage. In Table I the EMIRR is calculated at an EMI disturbance voltage of $100\,\text{mV}$ (peak voltage) for frequencies, that are commonly used for communication purposes as suggested in [14, p. 2].

Structure B, utilising RC low-pass filters, performs best in terms of EMIRR without changing the other transistor parameters significantly. A major drawback, however, is the significantly increased area requirement (about 8 times larger) for the whole amplifier circuit, assuming poly resistors and capacitances. The area demand is shown in Table I as a percentage of the reference circuit. It was calculated from the dimensioning of the transistors, resistors and capacitances given in the structures including the output branch. As mentioned previously the area can be reduced, but introducing the drawback of increased noise due to larger resistor sizes. The same problem also arises for structures F and G, that use similar dimensioning for the high-pass filters needed. Due to the filters only altering the circuit at higher frequencies (above the unity gain frequency), also the specifications given in section II for structure A are fully met. Similarly the increased area of structure H is depending on the capacitance C_H.

Structures C, D and E, that are adapting the concept of linearisation are much less area-demanding. However, these

also have a lower EMIRR. Furthermore, making the transconductance of the input pair constant over a wider range, comes with the negative effect of simultaneously decreasing the transconductance. Therefore the differential mode gain is decreased by a few dB. In structure E source degeneration and harmonics cancellation is applied. The combination of both linearisation techniques shows, that the more linearised the input stage is, the higher the EMIRR that can be reached, but also the lower the transconductance of the input stage becomes. Furthermore the GBWP is also decreased. The output swing and the power consumption are not significantly altered by any structure, but the phase margin might pose a problem for structure F, that utilises a cross-coupled differential pair with a high-pass filter.

Another important topic, that needs to be considered when using these structures in integrated circuits is the demand for matching or even absolute values. Starting with structure A, the input transistors themselves of course, need to match to the greatest extent possible. The same is true for structures B and C, and this not only for the input transistors, but also for the resistors and capacitances used. This is quite straightforward and therefore classified "easy". Structures D and E do not only demand matching two and two transistor pairs, but also the ratio between one and the other differential input pair must be adjusted precisely, requiring "medium" effort. On the contrary, structures F and H need four equally matched transistors in order to keep the offset voltage low. The greatest design efforts, however, must be concentrated on the structures G and H, since either the currents of the PMOS and NMOS transistors must be matched (structure G) or capacitances must be trimmed to exact values (structure H), leading to the classification "hard".

Besides investigating the EMIRR for specific frequencies, another analysis is shown in Figure 9, illustrating the EMI induced offset over a broader frequency range (from $1\,\text{MHz}$ to $10\,\text{GHz}$) at a peak amplitude of $500\,\text{mV}$. The behaviour of the different structures is shown as deviation from the technological offset. At frequencies up to $10\,\text{MHz}$ all structures behave similarly, as this still lies within the bandwidth of the amplifier. Due to the decreasing gain a negative offset voltage is induced by EMI. However, for higher frequencies the EMI countermeasures start inducing an inverse offset voltage. This works especially well for structures B, C, E, F, H. By contrast, structures D and G remain with larger offset voltages at high frequencies, whereas structure G performs worst around the corner frequency defined by the high-pass filter, when the NMOS differential pair is about to be turned on. Figure 9 shows, that structure G might not be the best choice even though it may have performed acceptably in Table I.

Outlook: The comparison of these different input pair structures was only dealt with on a single topology and dimensioning level as a case study. Further research will be put into the question, of whether different specifications and dimensioning of the nominal reference structure lead to the same qualitative behaviour of the EMI induced offset in respect to the EMI frequency. Additionally it is planned to implement these structures in terms of a test chip and to cross-check the simulated behaviour with the real behaviour.

TABLE I
SUMMARY OF FOLDED CASCODE STRUCTURES WITH DIFFERENT EMI ROBUST INPUT STAGES

Structure	Reference A	Filtering B	Linearisation			Compensation			
			C	D	E	F	G	H	
Technological DC Offset	61.77	61.77	200.5	92.29	165.5	82.37	106.8	61.77	µV
Differential Mode Gain	82.46	82.46	75.96	78.81	73.74	79.71	78.3	82.46	dB
Common Mode Gain	−21.5	−21.5	−9.24	−18.13	−17.2	−22.3	−21.6	−21.5	dB
GBWP	10.3	10.1	5.85	8.51	4.53	9.14	21.6	10.3	MHz
Phase Margin	86.0	72.7	86.7	85.4	86.7	57.3	79.7	86.0	°
Power Consumption	1.02	1.02	1.02	1.44	1.44	1.48	1.49	1.48	mW
Output Swing	2.65	2.65	2.6	2.54	2.52	2.63	2.65	2.65	V
Slew Rate Ratio (rising/falling)	48.3/51.7	49.5/50.5	50.2/49.8	48.6/51.4	49.6/50.4	50/50	53.9/46.1	49.0/51.0	%
Area Comparison	100	780	102	181	184	842	815	572	%
Matching/Trimming Effort	easy	easy	easy	medium	medium	easy	hard	hard	-
EMIRR @ 100 mV, 400 MHz	37.2	76.3	46.0	44.2	55.6	68.0	51.4	60.0	dB
EMIRR @ 100 mV, 900 MHz	38.5	93.0	47.6	45.6	61.0	75.3	47.7	59.5	dB
EMIRR @ 100 mV, 1.8 GHz	39.0	104.2	48.2	46.2	67.4	90.0	47.2	58.2	dB
EMIRR @ 100 mV, 2.4 GHz	39.1	109.5	48.3	46.1	69.5	80.6	47.3	59.3	dB

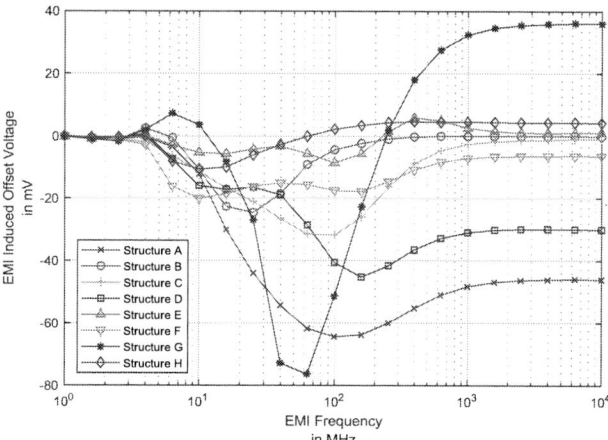

Fig. 9. EMI induced offset voltage over frequency for an EMI amplitude of 500 mV.

Figure 9 shows that the deviation in offset has different directions depending on the structure used. Therefore, it would also be of interest to investigate, whether it is possible to combine structures having a positive EMI induced offset, with structures having a negative EMI induced offset voltage. In addition it should be shown, that comparing the EMIRR on single frequencies may not be the best choice for classifying the EMI robustness of a structure.

V. CONCLUSION

In this paper various EMI robust differential input pair structures have been compared. The goal was to investigate the main advantages and disadvantages when applying different EMI mitigation techniques to a reference folded cascode operational amplifier. The measures applied were filtering, linearisation and compensation, whereas each of these offer well-functioning topologies. Special focus was put on how the specifications of the initial structure are affected by introducing the EMI countermeasures referred to. It was shown that the measures having the highest EMIRR are most area-consuming. By contrast linearisation techniques lead to a decreased differential mode gain. Therefore a trade-off between area, EMIRR and other amplifier specification values has to be made and this document should be a help for designers achieving this aim.

REFERENCES

[1] Jean-Michel Redoute and Anna Richelli. "A Fundamental Approach to EMI Resistant Folded Cascode Operational Amplifier Design". In: *2013 International Symposium on Electromagnetic Compatibility*. IEEE, 2013. ISBN: 9781467349802. URL: http://ieeexplore.ieee.org/servlet/opac?punumber=6636283.

[2] C. Walravens et al. "Efficient reduction of electromagnetic interference effects in operational amplifiers". In: *Electronics Letters* 43.2 (2007), p. 84. ISSN: 00135194. DOI: 10.1049/el:20073026.

[3] Farzan Rezaei and Seyed Javad Azhari. "Transconductor Linearization Based On Adaptive Biasing of Source-Degenerative MOS Transistors". In: *Circuits, Systems, and Signal Processing* 34.4 (2015), pp. 1149–1165. ISSN: 0278-081X. DOI: 10.1007/s00034-014-9902-6.

[4] Mohamed B. Elamien and Soliman A. Mahmoud. "A linear CMOS balanced output transconductor using double differential pair with source degeneration and adaptive biasing". In: *2016 IEEE 59th International Midwest Symposium on Circuits and Systems (MWSCAS)*. IEEE, 16.10.2016 - 19.10.2016, pp. 1–4. ISBN: 978-1-5090-0916-9. DOI: 10.1109/MWSCAS.2016.7870002.

[5] Franco Fiori. "Design of an Operational Amplifier Input Stage Immune to EMI". In: *IEEE Transactions on Electromagnetic Compatibility* 49.4 (2007), pp. 834–839. ISSN: 0018-9375. DOI: 10.1109/TEMC.2007.908255.

[6] P. S. Crovetti. "Operational amplifier immune to EMI with no baseband performance degradation". In: *Electronics Letters* 46.3 (2010), p. 207. ISSN: 00135194. DOI: 10.1049/el.2010.2853.

[7] J.-M. Redouté and M. Steyaert. "EMI resisting CMOS differential pair structure". In: *Electronics Letters* 42.21 (2006), p. 1217. ISSN: 00135194. DOI: 10.1049/el:20062059.

[8] Jean-Michel Redouté and Michiel Steyaert. *EMC of Analog Integrated Circuits*. Springer, New York, 2010. ISBN: 978-90-481-3229-4. DOI: 10.1007/978-90-481-3230-0. URL: http://dx.doi.org/10.1007/978-90-481-3230-0.

[9] Andrea Lavarda. "EMI susceptibility of offset compensated CMOS operational amplifiers". PhD thesis. Graz University of Technology, 2018. URL: https://permalink.obvsg.at/tug/AC15358154 (visited on 04/01/2020).

[10] Franco Fiori. "An Analog Front End Based on Chopped Signals Highly Immune to RFI". In: *2015 Asia-Pacific Symposium on Electromagnetic Compatibility (APEMC)*. IEEE, 2015, pp. 98–101. ISBN: 978-1-4799-6668-4. DOI: 10.1109/APEMC.2015.7175381.

[11] Texas Instruments. *EMI Rejection Ratio of Operational Amplifiers*. 2011. URL: https://www.ti.com/lit/an/sboa128a/sboa128a.pdf (visited on 03/23/2020).

[12] Gunter Winkler and Bernd Deutschmann. *EMI Rejection of Operational Amplifiers*. 2016. URL: http://www.seibersdorflaboratories.at/fileadmin/uploads/intranet/events/emv-nfc-ft/emv-vortr-2016/02_tugraz_winkler-deutschmann_emi_rejection.pdf (visited on 03/23/2020).

[13] Marco Brignone Aimonetto and Franco Fiori. "On the effectiveness of EMIRR to qualify OpAmp". In: *2015 IEEE International Symposium on Electromagnetic Compatibility (EMC)*. IEEE, 2015, pp. 40–44. ISBN: 978-1-4799-6616-5. DOI: 10.1109/ISEMC.2015.7256129.

[14] Texas Instruments. *OPA1662 EMI Immunity Performance*. 2012. URL: https://www.ti.com/lit/an/sbot006/sbot006.pdf (visited on 07/29/2020).

Integrated High Voltage Active Quenching Circuit in 150nm CMOS Technology

Martin Jungwirth
Inst. of Electrodynamics, Microwave and Circuit Engineering
TU Wien
Vienna, Austria
martin.jungwirth@tuwien.ac.at

Alija Dervić
Inst. of Electrodynamics, Microwave and Circuit Engineering
TU Wien
Vienna, Austria
alija.dervic@tuwien.ac.at

Horst Zimmermann
Inst. of Electrodynamics, Microwave and Circuit Engineering
TU Wien
Vienna, Austria
horst.zimmermann@tuwien.ac.at

Abstract—An integrated high voltage active quenching circuit (AQC) in a 150 nm high voltage CMOS technology is presented. The circuit is designed for off-chip single-photon avalanche diodes (SPAD) with parasitic capacitance up to 5 pF. With the high voltage lateral double-diffused MOSFETs (LDMOS) the circuit is able to apply excess bias voltages up to 35 V. The excess bias voltage for the SPAD is adjustable from 5 V up to 35 V and retains a constant quenching time of 2.2 ns at a SPAD capacitance of 5 pF.

Keywords—*SPAD, CMOS, high voltage, active quenching circuit, AQC, quenching circuit*

I. Introduction

Detecting faint optical signals is an important capability in many fields like telecommunication, quantum information science, quantum encryption, medical imaging and ranging. Single-photon avalanche diodes (SPADs) are broadly used detectors for single photons, replacing photomultiplier tubes (PMT), previously used for such applications [1]. SPADs are bringing advantages in terms of size, cost, required operation voltage, and quantum efficiency and are insensitive to magnetic fields.

The photon-detection probability (PDP) describes how efficiently photons are detected by the SPAD and can be expressed as

$$PDP(\lambda, V_{EX}) = \eta(\lambda) \cdot P_T(V_{EX}). \quad (1)$$

Where $\eta(\lambda)$ is the wavelength dependent photon absorption efficiency and $P_T(V_{EX})$ the probability of triggering an avalanche breakdown. This triggering probability increases with an increasing excess bias voltage V_{EX} as shown in [2]–[4]. Some characteristics and non-ideal effects of SPADs are important for the design of a quenching circuit. The first one is its parasitic capacitance C_{SPAD}, because it influences the amount of avalanche charge during a breakdown [1]. An important non-ideal effect of SPADs is the so-called afterpulsing effect. Afterpulsing describes additional avalanche breakdowns caused by trapped charges after an initial photon triggered avalanche breakdown.

In order to utilize the effect of increased *PDP* with higher V_{EX}, this paper presents an integrated high voltage active quenching circuit (AQC) for off-chip SPADs. The circuit is designed in a 150 *nm* CMOS process and uses lateral double-diffused MOSFETs (LDMOS), which are capable of switching up to 40 *V*. With this LDMOS it is possible to apply quenching voltages up to 35 *V* onto the SPAD and quench an avalanche in 2.2 ns.

II. Circuit Description

To achieve fast quenching the initial structure of the proposed circuit shown in Fig. 1 was chosen.

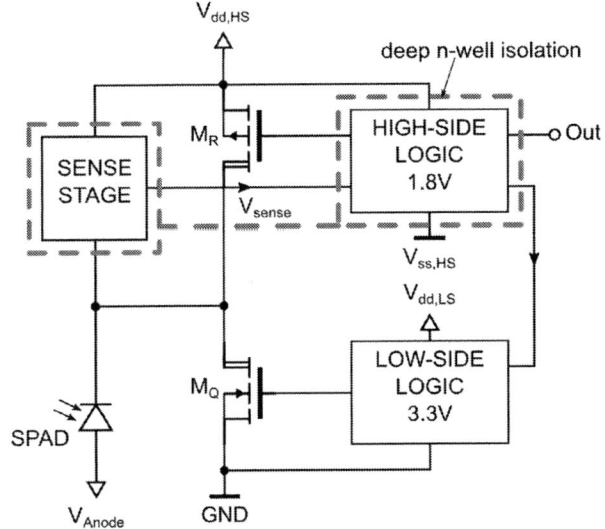

Fig. 1   Initial structure of the proposed circuit

This structure is based on the topology by Acconcia et al. [5] with major improvements in terms of the MOSFET type used as quenching switch. By using an n-channel LDMOS as quenching transistor instead of a p-channel transistor, the higher mobility of electrons compared to holes can be exploited to speed up the quenching time.

The circuit for the quencher is split up into two parts, the first one using 1.8 *V* transistors and the second one using 3.3 *V* transistors. The control logic in the high-side logic block and the sense stage, are designed with 1.8 *V* logic level transistors and placed in a deep n-well for isolation. The deep n-well isolation is necessary because the V_{SS}-potential is referenced to 35 *V* – 1.8 *V* against the substrate. By using the 1.8 *V* logic level transistor for the sense stage and control logic their speed can be exploited to reduce the quenching time. The low-side logic only contains the driver stage for the high voltage LDMOS M_Q and the low-side part of the required level shifter. This low-side logic is built with 3.3 *V* logic level transistors because the nominal gate voltage of the LDMOS is 3.3 *V*.

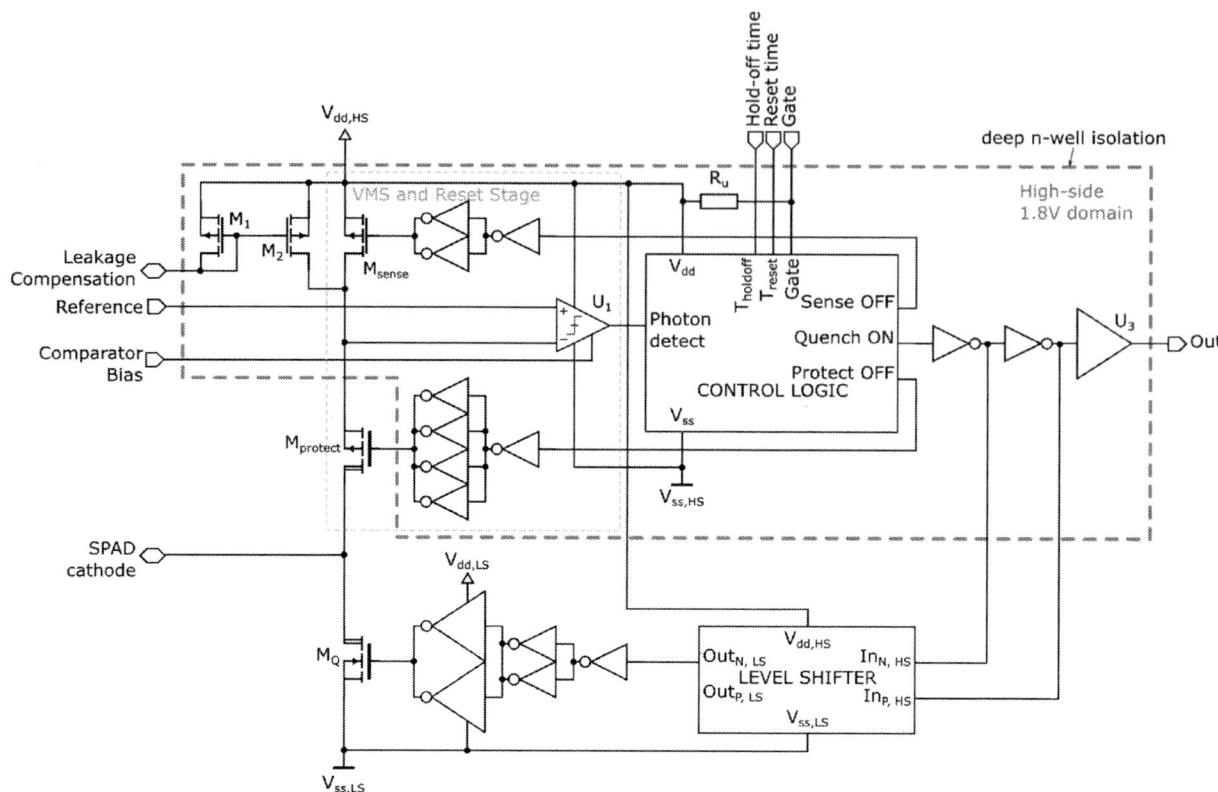

Fig. 2   Schematic of the high voltage active quenching circuit

The overall schematic of the proposed circuit is shown in Fig. 2 with the two high-voltage transistors M_Q, $M_{protect}$ and the low voltage transistor M_{sense} responsible for sensing, quenching, and resetting the SPAD. The LDMOS M_Q is the largest transistor in the design and required for fast quenching of an avalanche. Fast quenching of an avalanche is especially important when using higher excess bias voltages than in usual low voltage quenching circuits. The higher excess bias voltage inflicts a potentially larger avalanche charge. Thus, it is important to sense and quench the avalanche as fast as possible to prevent that the whole charge $Q_{EX} = C_{SPAD} \, V_{EX}$ stored in C_{SPAD} will flow through the space-charge region during an avalanche. Keeping the avalanche charge low helps in reducing the afterpulsing probability (APP).

A. Combined Sense Stage and Reset Switch

The sense circuit required for the detection of an avalanche is typically in parallel to the reset transistor as can be seen also in [5], [6]. To charge up C_{SPAD} of the off-chip SPAD, which is in the range of $5 \, pF$ a rather large reset transistor M_R is required. This large reset transistor introduces a large parasitic capacitance, which shunts the avalanche current away from the sense stage during the onset of an avalanche, thereby reducing the sensitivity to an avalanche. To avoid this effect the sense stage is combined with the reset switch using the transistors M_{sense} and $M_{protect}$ in Fig. 2. When the circuit is in ready state and waiting for an avalanche to detect, the logic level (1.8V) transistor M_{sense} is switched off, thereby imposing a high impedance to the SPAD-cathode through the

switched on LDMOS $M_{protect}$. Now the OFF-resistance of M_{sense} acts as current sense resistor for the voltage comparator U_1. At the onset of an avalanche breakdown, the avalanche current causes a voltage drop on the off-impedance of M_{sense}, which is detected by the voltage comparator U_1. Immediately after the detection of an avalanche the control logic switches off $M_{protect}$ and switches on M_{sense}. By switching off $M_{protect}$, the input to the comparator and M_{sense} are protected from the high quenching voltage step during the following quenching phase. Additionally, the switched-on M_{sense} connects the input of the comparator via a low impedance path to $V_{dd,HS}$ to further protect it from high voltage spikes due to the parasitic capacitances of the LDMOS. After quenching and after the hold-off time is over, $M_{protect}$ gets switched on by the control logic. Now both transistors M_{sense} and $M_{protect}$ are switched on and form the reset switch to recharge the SPAD to the desired bias point for the next photon detection.

B. Level Shifter

The level shifter block in Fig. 2 is an important circuit block in the overall design, which has a major impact on the quenching time. It is necessary to translate the M_Q gate signal from a high side $1.8 \, V$ logic signal to a low side $3.3 \, V$ logic signal. The current literature offers a wide selection of level shifter designs. Under the limitations of the used technology and with the requirement of a fast response, the design from Declerq et al. [7] was chosen. The actual implementation of this level shifter circuit is shown in Fig. 3. With the used technology, this level shifter achieves a response time of

100 ps from high-side to low-side. This fast response time comes with the price of a rather high power consumption due to the fact that one high voltage transistor is always switched on. The static power consumption of the level shifter alone is around 90 mW and is dominating the overall power consumption. This level shifter has also the advantage that the voltage difference between high-side and low-side can be freely chosen in the limits of the high voltage LDMOS. This means the applied quenching voltage to the SPAD (i. e. its excess bias) can be adjusted.

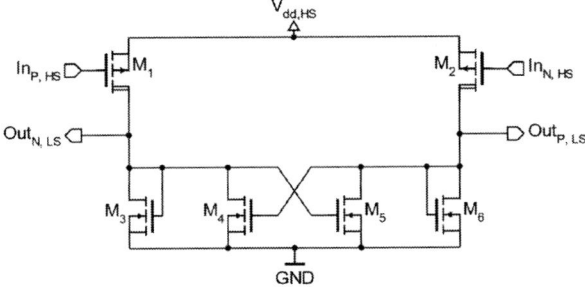

Fig. 3   Schematic of the implemented level shifter.

C. Comparator

For the detection of the avalanche, a fast, continuous-time comparator circuit is used. This comparator is based on a basic differential amplifier with positive feedback in the active load to enhance its switching speed. Followed by a push-pull output stage. The trigger level for this quenching circuit is a few 100 mV below the positive supply rail, to further reduce the quenching time. Thus, the comparator was optimized for a fast response at input voltages close to the positive supply rail.

D. Control Logic

The gate signals to control the three main transistors are generated in the control logic block. Therefore, the control logic has to react fast, when the comparator detects an avalanche and generate the gate signals in the right sequence. To achieve a fast response time, the *photon detect* input signal is fed forward and OR connected to the gate signals *Quench*

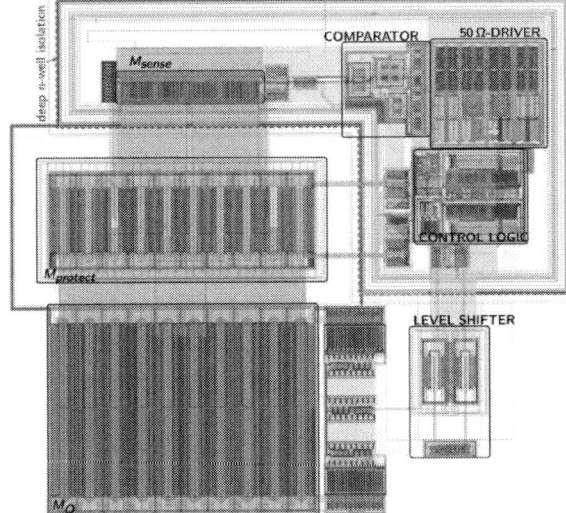

Fig. 4   Layout image of the high-voltage AQC.

ON and *Protect OFF*. Thereby bypassing the core logic, which generates the hold-off time and reset time for the quenching and reset phase. After the initial fast response to an avalanche, the adjustable pulse generators in the control logic take over the control of the gate signals. One pulse generator keeps the control logic in quenching state for the duration of the off-chip adjustable hold-off time. After the hold-off time is over, a second adjustable pulse generator ensures that for the duration of the reset time the corresponding gate signals are produced. An additional delay is inserted into the gate signal of $M_{protect}$ to prevent cross conduction with M_Q when switching from quenching into reset state. This delay is necessary due to the response time of the level shifter. After the reset time is over, the SPAD is recharged to its bias point and the control logic is ready again to detect another photon.

E. Layout

The layout for the designed high voltage AQC is shown in Fig. 4, with major components labelled. The dimensions of the circuit are 200 $\mu m \times$ 200 μm. Due to the high quenching voltages involved, all the logic circuits except the level shifter are embedded in a deep n-well isolation. This isolated circuit block, which generates the gate signals for the main transistors is closely fitted to $M_{protect}$ to reduce circuit parasitics.

III. SIMULATION RESULTS

Post-layout simulation results for a typical quenching sequence of the designed high voltage AQC are shown in Fig. 5. For these simulations, the excess bias voltage was set to 35 V and the reference voltage for the comparator was set to 100 mV below $V_{dd,HS}$. For simulating the SPAD a simplified version of the model in [8] was used. The simplification was to replace the piecewise linear voltage source with a DC voltage source for the breakdown voltage of the SPAD and a resistor. The resistor is used to model the slope of the $I - V$ characteristic during breakdown. For the results in Fig. 5 the

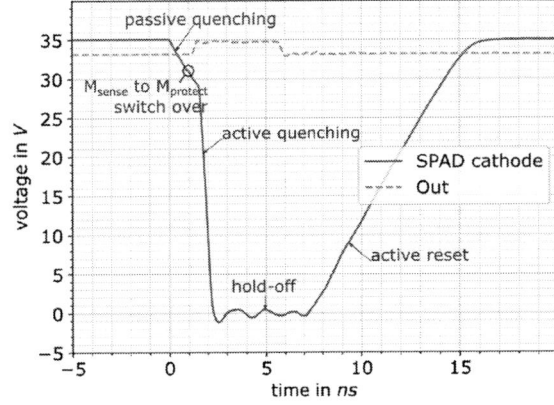

Fig. 5   Post-layout simulated SPAD-cathode voltage and output signal.

SPAD capacitance was 5 pF and the intrinsic resistance was modelled to be 800 Ω according to [9]. The blue graph shows the voltage at the SPAD cathode during quenching and the orange graph shows the corresponding output signal of the AQC. Immediately after triggering of an avalanche, passive quenching starts due to the high OFF-resistance of M_{sense}. At around 800 ps the comparator has already detected the

978-1-7281-8494-4/20 $31.00 © 2020 IEEE

avalanche and has generated the required gate signals for M_{sense} and $M_{protect}$. This can be seen at the point marked "M_{sense} to $M_{protect}$ switch over" where $M_{protect}$ is switched off and nearly at the same time M_{sense} is switched on. This is done to protect M_{sense} and the comparator from the high quenching voltage. At around $1.5\,ns$ the transistor M_Q is switched on to rapidly discharge the SPAD and lower its bias voltage below the breakdown voltage. The steep falling edge during active quenching lasts $0.7\,ns$ and therefore the total quenching time is 2.2ns. For the duration of the adjustable hold-off time, the SPAD bias voltage is kept below the breakdown voltage. Ensuring a hold-off time after quenching an avalanche, is another countermeasure to afterpulsing. Following the hold-off time, the reset phase takes place, where the SPAD is recharged to its initial bias point. The reset time is $9.6\,ns$, which is considerably longer compared to the quenching time. The longer reset time was chosen intentionally, in favour of a short quenching time.

The behaviour of the circuit for different excess bias voltages is shown in Fig. 6. Changing the excess bias voltage is possible because of the used level shifter in the circuit. It can be seen that the quenching time is nearly identical for the different voltages. Whereas the reset time depends on the used excess bias voltage of the used SPAD. To accommodate for different excess bias voltages and SPAD capacitances the reset time is also adjustable.

The designed AQC offers the possibility to adjust the hold-off time between $5\,ns$ up to $160\,ns$. Thereby quenching time and reset time, are independent from the chosen hold-off time.

In Table 1 the characteristic properties of recent state-of-the-art high voltage AQC are compared with the presented design. It should be noted that the parameters from the referenced designs are taken from measurements, whereas in this paper post-layout simulation data is available for the presented design. It can be seen that this design achieves a similar maximum count rate compared to [5] at twice the excess bias voltage. It is also worth noticing that in [5] it is stated that active quenching starts $2.3\,ns$ after an avalanche is detected, whereas in this design an avalanche is already completely quenched in $2.2\,ns$, which is a similar time as the recent design in [3] which has a considerably lower maximum quenching voltage of $9.9\,V$. Reducing the quenching time is crucial for reducing the afterpulsing probability.

TABLE 1 COMPARISON OF AQC PARAMETERS

design	max. quenching Voltage (V)	photon count rate (Mcounts/s)	quenching time (ns)	reset time (ns)
this design	35	90 @ 10 V	2.2	9.6
[5]	50	80 @ 5 V	60 @ 30 V	40 @ 30 V
[6]	20	20	25	20
[3]	9.9	129	2.2	2.4

IV. CONCLUSION

We present a design for a high voltage active quenching circuit for the use with off-chip SPADs in a $150\,nm$ high voltage CMOS technology. The designed AQC is able to apply excess bias voltages up to $35\,V$ onto the SPAD, by using high voltage lateral double-diffused MOSFETs (LDMOS). With the switching speed of this LDMOS, the circuit quenches an avalanche in $2.2\,ns$ at a SPAD capacitance of $5\,pF$. At an excess bias voltage of $10\,V$ a photon count rate of $90\,Mcounts/s$ is achieved. The presented design will therefore allow large excess bias voltages with short quenching time, potentially resulting in high PDP and low APP.

ACKNOWLEDGMENT

The authors acknowledge financial support from the Austrian Science Fund (FWF, grant no. P28335-N30). M. J. thanks M. Hofbauer for helpful discussions.

REFERENCES

[1] A. Gallivanoni, I. Rech, and M. Ghioni, "Progress in Quenching Circuits for Single Photon Avalanche Diodes," *IEEE Trans. Nucl. Sci.*, vol. 57, no. 6, pp. 3815–3826, Dec. 2010.

[2] F. Zappa, S. Tisa, A. Tosi, and S. Cova, "Principles and features of single-photon avalanche diode arrays," *Sens. Actuators Phys.*, vol. 140, no. 1, pp. 103–112, Oct. 2007, doi: 10.1016/j.sna.2007.06.021.

[3] A. Dervić, B. Steindl, M. Hofbauer, and H. Zimmermann, "High-voltage active quenching and resetting circuit for SPADs in 0.35 μm CMOS for raising the photon detection probability," *Opt. Eng.*, vol. 58, no. 4, p. 040501, Apr. 2019, doi: 10.1117/1.OE.58.4.040501.

[4] M. Ghioni, S. Cova, F. Zappa, and C. Samori, "Compact active quenching circuit for fast photon counting with avalanche photodiodes," *Rev. Sci. Instrum.*, vol. 67, no. 10, pp. 3440–3448, Oct. 1996, doi: 10.1063/1.1147156.

[5] G. Acconcia, I. Rech, A. Gulinatti, and M. Ghioni, "High-voltage integrated active quenching circuit for single photon count rate up to 80 Mcounts/s," *Opt. Express*, vol. 24, no. 16, pp. 17819–17831, Jul. 2016, doi: 10.1364/OE.24.017819.

[6] F. Zappa, A. Lotito, A. C. Giudice, S. Cova, and M. Ghioni, "Monolithic active-quenching and active-reset circuit for single-photon avalanche detectors," *IEEE J. Solid-State Circuits*, vol. 38, no. 7, pp. 1298–1301, Jul. 2003, doi: 10.1109/JSSC.2003.813291.

[7] M. J. Declerq, M. Schubert, and F. Clement, "5 V-to-75 V CMOS output interface circuits," in *1993 IEEE International Solid-State Circuits Conference Digest of Technical Papers*, Feb. 1993, pp. 162–163, doi: 10.1109/ISSCC.1993.280014.

[8] A. D. Mora, A. Tosi, S. Tisa, and F. Zappa, "Single-Photon Avalanche Diode Model for Circuit Simulations," *IEEE Photonics Technol. Lett.*, vol. 19, no. 23, pp. 1922–1924, Dec. 2007.

[9] M. Stipčević, H. Skenderović, and D. Gracin, "Characterization of a novel avalanche photodiode for single photon detection in VIS-NIR range," *Opt. Express*, vol. 18, no. 16, pp. 17448–17459, Aug. 2010.

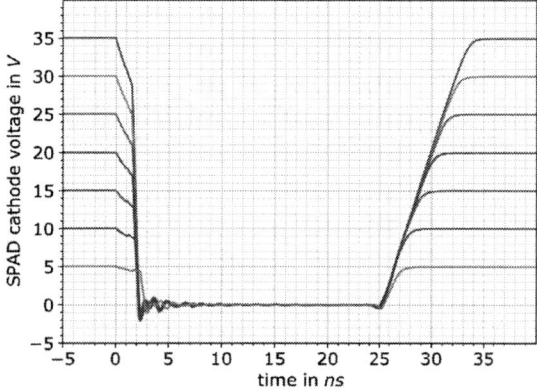

Fig. 6 SPAD cathode voltage transients for different excess bias voltages.

Low Power Ku- and Ka-Band SiGe HBT Low-Noise Amplifiers

Syed Sharfuddin Ahmed
Institute of Electron Devices and Circuits
Ulm University
89081 Ulm, Germany
syed.ahmed@uni-ulm.de

Hermann Schumacher
Institute of Electron Devices and Circuits
Ulm University
89081 Ulm, Germany
hermann.schumacher@uni-ulm.de

Abstract—This paper presents low power, high gain, low-noise amplifiers (LNA) at Ku- and Ka-band frequencies in a 0.13μm SiGe:C BiCMOS technology. The designed LNAs fulfill the RF performance requirements of active phased-array antennas for satellite communications or 5G applications while consuming less than 6 mW of power. The Ku-band LNA exhibits a peak gain of 28 dB with a 3-dB bandwidth (BW) of 3.6 GHz (9.2 – 12.8 GHz), noise figure (NF) of <1.9 dB and IP1dB of -30 dBm. The Ka-band LNA bandwidth extends between 18-21.1 GHz while providing 26 dB of peak gain with a noise figure of <2.3 dB and IP1dB of -30 dBm. Among published LNAs in Silicon technologies, the designed LNAs demonstrate the best figure of merit, considering noise, gain, linearity, and consumed power.

Index Terms—Ku-band; Ka-band; low-noise amplifier (LNA); satellite communications; SiGe HBT; phased-array; low power

I. INTRODUCTION

Mobile satellite-enabled internet services (Satcom-on-the-move) enable high data rate services with an unparalleled degree of flexibility and mobility. Recently, electronically steerable phased-array antennas have become a prime research area for this purpose. Due to the close proximity of antenna elements at microwave frequencies above 10 GHz, such systems need to be realized using high-performance microwave/millimeter-wave integrated circuits (MMICs) in the transmit/receive (T/R) modules. Achieving the challenging EIRP and G/T specifications of satellite systems requires hundreds, if not thousands of T/R modules, making their individual power consumption a critical challenge. The low noise amplifier (LNA) determines the receiver sensitivity, and, due to linearity requirements connected to in-array cross-talk, its power consumption can be a significant contribution to the overall dissipation.

This work presents two LNAs in Ku- (9.2-12.8 GHz) and Ka-band(18-21.1 GHz), designed to achieve low dc power dissipation while maintaining desired gain and minimizing noise figure (NF). The LNAs were designed in low cost, high performance 0.13μm SiGe:C BiCMOS technology having transit frequency f_t of 250 GHz, and oscillation frequency f_{max} of 300 GHz. The process offers two thick metal layers with relatively low sheet resistance and five thin metal layers. In the following section, design steps for both the LNA are

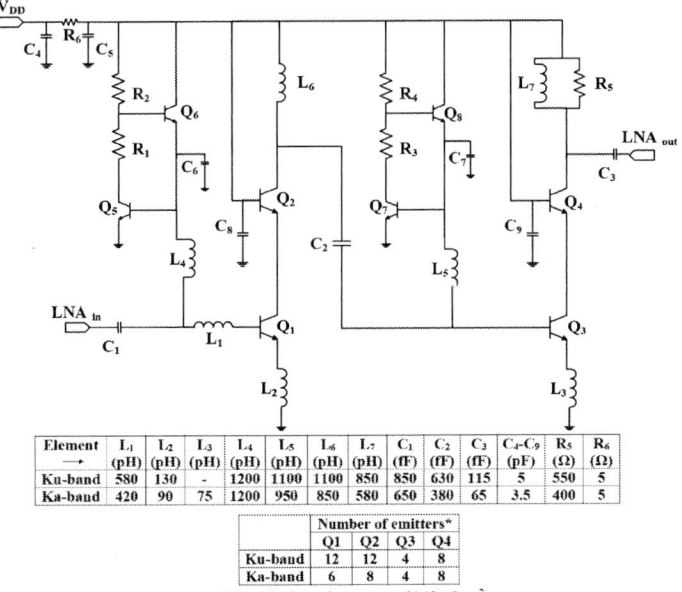

Element	L_1 (pH)	L_2 (pH)	L_3 (pH)	L_4 (pH)	L_5 (pH)	L_6 (pH)	L_7 (pH)	C_1 (fF)	C_2 (fF)	C_3 (fF)	C_4-C_9 (pF)	R_5 (Ω)	R_6 (Ω)
Ku-band	580	130	-	1200	1100	1100	850	850	630	115	5	550	5
Ka-band	420	90	75	1200	950	850	580	650	380	65	3.5	400	5

	Number of emitters*			
	Q1	Q2	Q3	Q4
Ku-band	12	12	4	8
Ka-band	6	8	4	8

*all emitter fingers have an area of 0.12×2 μm^2

Fig. 1: Schematic and component values of the Ku-band and Ka-band LNA

explained and later post-layout simulation results are presented in section III.

II. CIRCUIT DESIGN

The proposed Ku-band and Ka-band LNAs are designed as two cascode stages where the first stage is optimized for noise and the second stage for gain/linearity while maintaining low power consumption. The optimization strategy followed in this work to have low power while achieving the required gain and noise figure (NF) is described in detail in the next section. The circuit architectures of both LNA are identical, but the element values are different for each band, as shown in Fig. 1.

The base and emitter inductors L_1 and L_2 from the first stage of the LNA are the part of the input matching network. The base inductor L_1 resonates the base emitter capacitance of input transistor Q_1 while the emitter degenerative inductor

L_2 modifies the real part of the base input impedance, thus achieving simultaneous noise and power match. Inductor L_5 and L_6 resonate the base capacitance Q_3 and collector capacitance of Q_2, respectively. Thus, the load of the first stage is formed by the real part of the second stage's input impedance parallel with the first stage output resistance. Inductor L_7 and capacitor C_3 form the output matching network. Resistor R_5 extends the bandwidth of the LNA.

The size of the second stage cascode transistor Q_4 is twice that of the lower transistor Q_3, which leads to a smaller V_{be} drop in Q_4 resulting in increased headroom for Q_3, thus improving the linearity. In the Ka-band LNA, the emitter degeneration inductance L_3 is used at the second stage of the LNA to further increase the linearity.

The LNAs were biased by a supply voltage V_{DD} of 1.6 V. Resistors R_1 to R_4 along with transistors Q_5 to Q_8 form the biasing network. Both LNA stages were biased with a current mirror with base current compensation technique via transistor Q_6 and Q_8. Inductor L_4 was used to bias the input transistor instead of a resistor to reduce the noise contribution. It is designed as high value as possible while maintaining a safe margin to self-resonance frequency and considering the area constraints. High-value capacitors C_4 to C_9 are implemented to provide RF grounds. Post-layout simulation for the Ku-band LNA showed some possible resonance of those shunt capacitors with the supply line's inductance, thus a resistance of small value R_6 was added to damp out any resonance.

A. Design for low noise and low power

In a conventional LNA design approach, the optimum collector current density is set for the lowest noise figure NF_{min}. Then the size of the input transistor is scaled to achieve an optimum noise source resistance R_{sopt} close to the real part of the system impedance, which is typically 50Ω [1]. In this work, the LNA design is performed similarly, with slight modifications according to the low power design approach presented in [2] [3].

As per the conventional LNA design approach, initially, for both of the LNAs, the optimum collector current density J_{copt} was determined for the minimum noise figure NF_{min} of the input transistor Q_1, having a unit cell transistor size of $l_e = 0.12 \times 2\mu m^2$. Then scaling of the transistor size l_e is performed while keeping J_{copt} constant. Transistor scaling was done by changing the number of emitter N_E while keeping the emitter size the same. Fig. 2a and 2b show the effect of the transistor scaling on R_{sopt} while maintaining the determined J_{copt} for Ku-band and Ka-band LNA respectively.

For the Ku-band LNA, from Fig. 2a. it can be seen that for the transistor size of $l_e = 14 \times (0.12 \times 2\mu m^2)$, R_{sopt} approaches 50Ω while the collector current I_C is 5.2 mA. In this design, to facilitate low power consumption, a slightly lower $l_e = 12 \times (0.12 \times 2\mu m^2)$ was chosen which reduces the I_C to 4.2 mA but increases R_{sopt} to 62Ω.

For the Ka-band LNA, as shown in Fig. 2b, a slightly reduced transistor size of $l_e = 6 \times (0.12 \times 2\mu m^2)$ was chosen,

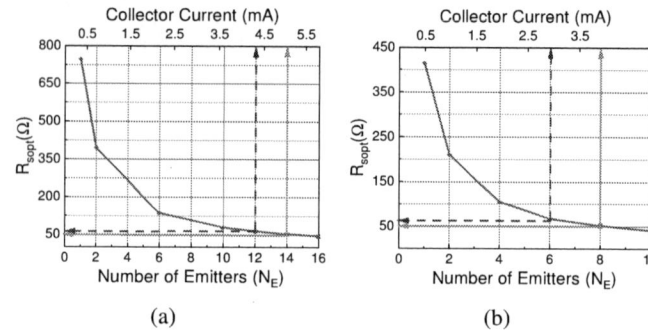

Fig. 2: Simulated R_{sopt} vs l_e biased at J_{copt} (a) Ku-band (b) Ka-band

reducing I_C from 3.7 mA to 2.8 mA, but increasing R_{sopt} from 50Ω to 65Ω.

For a fixed transistor size of l_e, modifying the collector current changes R_{sopt} [2] [3]. To further lower the collector current I_c and to check its effect on R_{sopt} and NF_{min} Fig. 3 is used for transistor size $l_e = 12 \times (0.12 \times 2\mu m^2)$ (for Ku-band), and $l_e = 6 \times (0.12 \times 2\mu m^2)$ (for Ka-band). Typically the system impedance of the transceiver module is 50Ω and it is evident from Fig. 3 that selecting collector current I_C lower than I_{Copt} not only reduces power consumption but also facilitates 50Ω match at the input, having only minor degradation of the minimum noise figure.

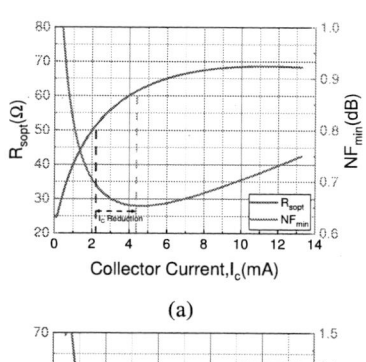

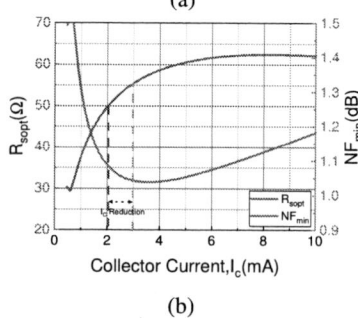

Fig. 3: Simulated R_{sopt} and NF_{min} vs I_C for a fixed l_e (a) Ku-band (b) Ka-band

From Fig. 3a, it can be seen for the Ku-band LNA, the first stage collector current is chosen 2.28 mA rather than 5.2 mA, found initially in the conventional approach. Thus 56% reduction of the power consumption is achieved with a slight increase in minimum noise figure from 0.63 dB to 0.7 dB.

Similarly, for the Ka-band LNA, the collector current of 2.2 mA is selected for the first stage, thus reducing the power consumption from the conventional approach by 40% while increasing the minimum noise figure by 0.05dB.

Now after setting the input transistor size and selecting the desired collector current, power consumption can further be decreased by reducing the supply voltage. Operating the transistor in the saturation region is avoided as in this region transistor base is flooded with excess minority carrier due to forward biasing of the collector-base junction resulting in performance degradation. As mentioned above, the designed LNA has a cascode architecture where the base of the cascode transistor is tied to the supply voltage V_{DD}. So, the collector-emitter voltage (V_{ce}) of the lower transistor will be lower than the supply voltage V_{DD} by V_{be}, (V_{DD}-V_{be}). The value of V_{be} for input transistor Q_1 is already set to be 0.79 V for Ku-band LNA and 0.78 V for Ka-band LNA by the choice of collector current. To lower the power consumption while providing enough headroom for the transistors a supply voltage of 1.6 V is chosen for both the LNAs.

The second stage is optimized for providing the desired gain while keeping the current consumption low. The impact of the noise figure of the second stage is small as its noise will be divided by the gain of the first stage. To set up the bias point of the second stage input transistor Q_3, Fig.4 is used. Fig. 4a and Fig. 4b displays the maximum available gain (G_{max}), NF_{min}, and the collector current density (J_c) vs the base-emitter voltage (V_{be}) for Ku-and Ka-band respectively. The NF_{min} has its minimum, around 0.825 V (Ku-band) and 0.83V (Ka-band) while the maximum of G_{max} occurs at 0.96 V (for both bands). Increasing V_{be} above 0.96 V would result in the transistor going deep into saturation thus G_{max} decreases rapidly while NF_{min} increases significantly. Considering the power consumption, gain and the noise figure requirements, for the second stage input transistor, V_{be} of 0.81 V (for both band) is chosen.

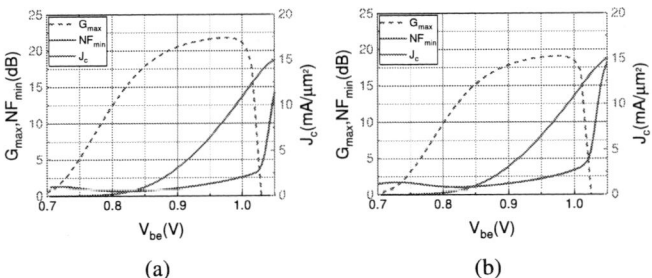

Fig. 4: Simulated G_{max}, NF_{min}, and J_C vs base-emitter voltage (V_{be}) at (a) Ku-band (11 GHz) (b) Ka-band (20 GHz)

B. Layout

To maximize the performance of the simulated LNA, careful layout optimization was done to decrease the parasitic effects and also reduce the total chip area. The inductor is the most area consuming component in the layout. So, to reduce the area of the total LNA chip, inductors over 500 pH were realized by stacking the top two metal layers, which have the lowest sheet resistance among the seven metal layers available in the technology. The other inductors were realized with the top metal layer. To avoid on-chip parasitic coupling effect inductors L_1, L_4, L_5-L_7 were bounded by a shield made by stacking all the metal layers in the technology. To further improve the noise performance, base inductor L_1 and emitter degeneration inductor L_2 is designed to have a high-quality factor (Q). The quality factor of these two inductors is 14 while other inductors have a Q between 9 to 11 for both the band.

In the LNA design, due to process constraints, realizing transistors with N_E over 8 is not possible with a single transistor. So, multiple transistors in parallel were used. To connect these transistors routing was done carefully. As the base is the most critical for noise performance, base routing was done with the top metal layer to reduce parasitic base resistance. In general, to reduce the interconnect parasitics, routing was performed as much as possible using the top two metal layers.

Using the Cadence RC parasitic extraction tool, influences of the parasitic capacitance especially at the high impedance such as first and second stage output nodes were examined. Most of the interconnects were fully EM simulated using ADS Momentum and the designed inductors were modified accordingly considering the parasitics.

The base of the cascode transistors Q_2 and Q_4 needed to have good RF ground thus long connections between the base and the MIM capacitor were avoided. Parasitic inductance created due to large connections may lead to instability. Also a large number of vias were used to connect to the ground plane. To ensure good RF grounds for the supply and bias voltages, capacitances were distributed throughout the layout. The dc connections were routed using the two top metal layers to reduce the series impedance. The bottom metal layer Metal 1 was used for routing the ground.

III. SIMULATION RESULTS

The initial design and simulation of the LNAs were performed using Keysight ADS software. Then inductors were designed and EM simulated using ADS Momentum. The layout was performed in Cadence Virtuoso and the Assura QRC tool was used for the parasitic extraction. Also, the layout interconnects were EM simulated with ADS Momentum. The circuit components were retuned and layout was optimized after observing the effect of parasitics from the layout and considering the parasitic effects of pads and bond-wires. Finally, the post-layout simulation was performed and the results are presented here.

Fig. 5 shows the post-layout s-parameter simulation results for the Ku-band LNA. The circuit consumes 3.58 mA current from 1.6 V supply voltage thus making the total power consumption of 5.72 mW. Fig. 5a shows the simulated peak gain (S_{21}) of 28 dB at 11.75 GHz and 3dB bandwidth (BW)

978-1-7281-8494-4/20 $31.00 © 2020 IEEE

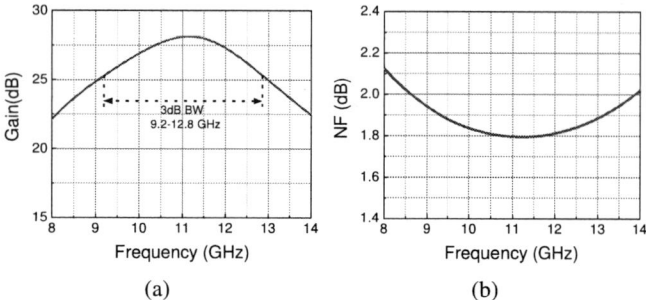

(a)   (b)

Fig. 5: Ku-band (a) Simulated Gain (b) NF vs Frequency

extends between 9.2 to 12.8 GHz, corresponding to a 33% relative bandwidth. The LNA exhibits a noise figure of 1.8 to 1.9 dB in the frequency range shown in Fig. 5b.

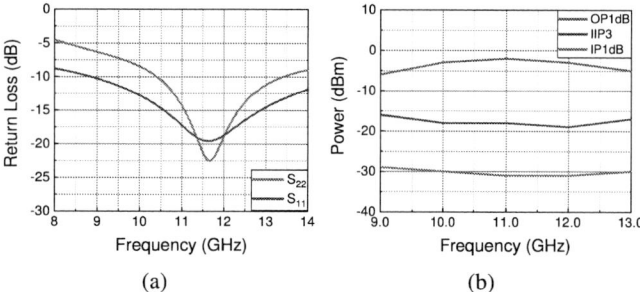

(a)   (b)

Fig. 6: Ku-band (a) Return Loss (b) 1-dB compression and third-order intercept point vs Frequency

The simulated input return loss (S_{11}) is more than -10 dB and output return loss (S_{22}) is more than -7dB along the frequency range, depicted in Fig. 6a.

The LNA linearity was checked by harmonic balance simulation and the results are summarized in Fig. 6b. The input-referred 1-dB compression point (IP1dB) is -29 to -31 dBm along with the BW. The third-order intercept point (IIP3) was simulated with two tone harmonic balance simulation with a 500 kHz tone separation. The obtained IIP3 is between -17.5 to -19 dBm.

The Ka-band post-layout simulation results are shown in Fig. 7 and 8. The LNA consumes a total of 5.88 mW power.

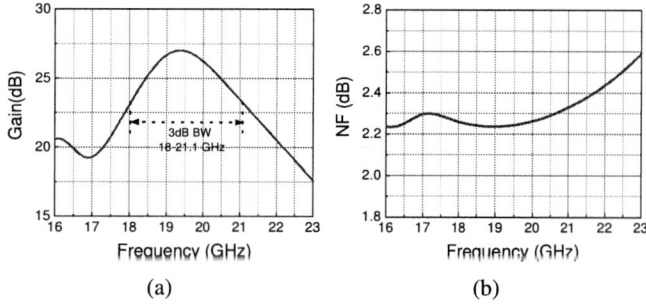

(a)   (b)

Fig. 7: Ka-band (a) Simulated Gain (b) NF vs Frequency

The LNA peak gain is about 27 dB and 3-dB bandwidth extends between 18 to 21.1 GHz. As shown in Fig. 7b in this frequency range LNA has a noise figure of 2.3 dB. The Ka-band LNA exhibits more than -20 dB input return loss (S_{11}) and more than -5 dB output return loss (S_{22}), showed in Fig 8a. Similar to Ku-band, Ka-band LNA linearity was also simulated. For the Ka-band LNA IP1dB of -28 to -30.5 dBm is obtained while IIP3 is between -17 to -20 dBm over the bandwidth, as shown in Fig. 8b.

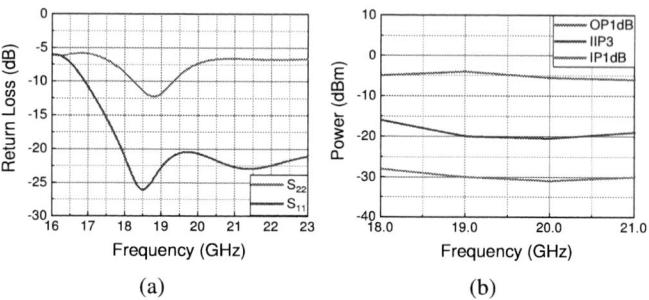

(a)   (b)

Fig. 8: Ka-band (a) Simulated Gain (b) NF vs Frequency

In the Fig. 9 layouts of both the LNA is shown. The Ku-band LNA has an area of $642 \times 360 \mu m^2$ and Ka-band LNA consumes $622 \times 338 \mu m^2$ of area. The pads are also included in the area calculation.

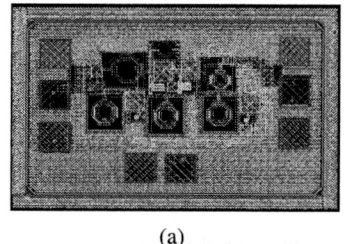

(a)

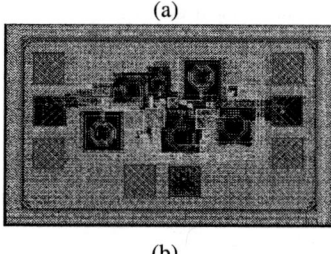

(b)

Fig. 9: Layout including pads (a) Ku-band (b)Ka-band

In order to compare the present work with other published state-of-the-art SiGe LNAs operating in the similar bands, a common benchmark is established via the LNA figures-of-merit (FOMs) shown in equation(1) [4]

$$FOM = \frac{S_{21}[abs].P_{1dB}[mW].Freq[GHz]}{(NF[abs] - 1).P_{dc}[mW]} \quad (1)$$

A higher FOM represents better performance. The comparisons between different published LNAs are summarized in Table I. As can be seen from Table I, both LNAs have a higher figure of merit, and are comparable with the state-of-art.

978-1-7281-8494-4/20 $31.00 © 2020 IEEE

TABLE I : Performance Summary and Comparison

		Frequency (GHz)	Gain (dB)	NF (dB)	IP1dB (dBm)	IIP3 (dBm)	Pdc (mW)	FOM	Technology
Ku-Band LNA	*This Work*	9.2-12.8	28	1.9	-29	-17.5	5.72	3.39	0.13 μm SiGe BiCMOS
	[1] 2006	8.25-13	11	2.78	-19.1	-9.1	2.5	0.7	180 GHz ft SiGe HBT
	[5] 2014	6.4-11	24.2	1.2	-19	-11	32.8	3.1	0.18 μm SiGe BiCMOS
	[6] 2014	10-22	15.5	3.4	-36.5	-33	4	0.02	0.13 μm SiGe BiCMOS
	[7] 2019	8-12	26	0.98	-17.3	-7.5	62	3.34	0.13 μm SiGe BiCMOS
Ka-Band LNA	*This Work*	18-21.1	27	2.3	-29	-17	5.88	2.78	0.13 μm SiGe BiCMOS
	[5] 2014	16-24	19	2.2	-16	-4	22.5	2.61	0.18 μm SiGe BiCMOS
	[8] 2017	19.25-21.5	32.8	3.5	-34.3	-21.2	7	1.55	0.25 μm SiGe BiCMOS
	[9] 2018	14.4-21.4	20.3	2.14	-23.7	-15.3	18	0.8	0.18 μm SiGe BiCMOS
	[10] 2018	19.3-27	12	2.6	-10	-1	27	1.34	0.13 μm SiGe BiCMOS

IV. CONCLUSION

This paper has demonstrated a low power, low noise, and moderate gain LNA in Ku- and Ka-band while maintaining desired linearity performance. The design methodology was explained briefly and the compromise between power consumption and other LNA performances such as gain, linearity, noise figure was discussed. The LNAs consume less than 6 mW of power while delivering peak gain of 27 dB and 28 dB at Ku-band and Ka-band respectively, along with a low noise figure, which makes them very suitable for large scale phased array application such as SatCom. Both LNAs have been submitted for fabrication. Due to careful optimization during and post layout, the measurement results are expected to be very similar to the post-layout simulation results presented here.

REFERENCES

[1] Kuo, W-ML, et al. "An X-band SiGe LNA with 1.36 dB mean noise figure for monolithic phased array transmit/receive radar modules." IEEE Radio Frequency Integrated Circuits (RFIC) Symposium, 2006. IEEE, 2006.

[2] Kuo, W-ML, et al. "A Low-Power, X-Band SiGe HBT Low-Noise Amplifier for Near-Space Radar Applications." IEEE Microwave and wireless components letters 16.9 (2006): 520-522.

[3] Thrivikraman, Tushar K., et al. "A 2 mW, sub-2 dB noise figure, SiGe low-noise amplifier for X-band high-altitude or space-based radar applications." 2007 IEEE Radio Frequency Integrated Circuits (RFIC) Symposium. IEEE, 2007.

[4] Lu, Chuang, et al. "A 20 GHz 1.9 dB NF LNA with distributed notch filtering for VSAT applications." 2014 IEEE MTT-S International Microwave Symposium (IMS2014). IEEE, 2014.

[5] Kanar, Tumay, and Gabriel M. Rebeiz. "X-and K-band SiGe HBT LNAs with 1.2-and 2.2-dB mean noise figures." IEEE transactions on microwave theory and techniques 62.10 (2014): 2381-2389.

[6] F. Inanlou, C. T. Coen and J. D. Cressler, "A 1.0 V, 10–22 GHz, 4 mW LNA Utilizing Weakly Saturated SiGe HBTs for Single-Chip, Low- Power, Remote Sensing Applications," IEEE Microw. and Wireless Compon. Lett., vol. 24, no. 12, pp. 890-892, Dec. 2014.

[7] Çaışkan, Can, et al. "Sub-1-dB and Wideband SiGe BiCMOS Low-Noise Amplifiers for X-Band Applications." IEEE Transactions on Circuits and Systems I: Regular Papers 66.4 (2018): 1419-1430.

[8] Tabarani, Filipe, and Hermann Schumacher. "A 32.8 dB gain, 3.5 dB NF, 7 mW, 20.35 GHz LNA with embedded 30 GHz band-stop filter." 2017 IEEE Bipolar/BiCMOS Circuits and Technology Meeting (BCTM). IEEE, 2017.

[9] Aljuhani, Abdurrahman H., Tumay Kanar, and Gabriel M. Rebeiz. "A Packaged Single-Ended K-Band SiGe LNA with 2.14 dB Mean Noise Figure." 2018 IEEE BiCMOS and Compound Semiconductor Integrated Circuits and Technology Symposium (BCICTS). IEEE, 2018.

[10] Issakov, Vadim, et al. "A 6 kV ESD-protected low-power 24 GHz LNA for radar applications in SiGe BiCMOS." 2018 IEEE BiCMOS and Compound Semiconductor Integrated Circuits and Technology Symposium (BCICTS). IEEE, 2018.

Fully Integrated Actively Quenched SPAD in 0.18µm CMOS Technology

Kerstin Schneider-Hornstein
Inst. of Electrodynamics, Microwave and Circuit Engineering
TU Wien
Vienna, Austria
Kerstin.Schneider-Hornstein@tuwien.ac.at

Michael Hofbauer
Inst. of Electrodynamics, Microwave and Circuit Engineering
TU Wien
Vienna, Austria
Michael.Hofbauer@tuwien.ac.at

Bernhard Steindl
Inst. of Electrodynamics, Microwave and Circuit Engineering
TU Wien
Vienna, Austria
Bernhard.Steindl@tuwien.ac.at

Horst Zimmermann
Inst. of Electrodynamics, Microwave and Circuit Engineering
TU Wien
Vienna, Austria
Horst.Zimmermann@tuwien.ac.at

Abstract—A fully integrated single- photon avalanche diode (SPAD) in a 180 nm high voltage CMOS technology is presented. The introduced quenching circuit is realized by the 3.3V high voltage transistors of the process to increase the excess bias voltage V_{ex} above the usual 1.8V supply voltage of the 180nm CMOS technology. Furthermore the circuit is cascoded to increase the excess bias even more up to 6.6V.

Keywords—SPAD, optical receiver, CMOS, active quenching

I. INTRODUCTION

The idea to detect less and less optical power drives the development of optical sensors and receivers towards the use of single-photon avalanche diodes (SPADs). These detectors are able to detect single photons hitting the active area. To achieve this, avalanche photodiodes (APDs) are biased in the so called Geiger mode. The bias voltage of the APD is increased beyond the breakdown voltage (V_{BR}) and as soon as a photon hits, the breakdown happens, an avalanche of charge carriers is built up and the subsequent voltage drop across the diode quenches the bias voltage below V_{BR} and therefore the avalanche is pinched off.

There are different types of quenching circuits (QCs), passive ones and active QCs. The simplest passive quencher consists of a resistor. This circuit has the disadvantage that the quenching time is rather long and the following reset time until the bias voltage (V_{bias}) is again set above V_{BR} is even longer. V_{bias} is defined by V_{BR} and the excess voltage V_{EX} ($V_{bias} = V_{BR} + V_{EX}$). To increase the speed of a quenching cycle, active quenchers detect the voltage drop across the resistor, decrease V_{bias} rapidly below V_{BR} and set it back to $V_{BR} + V_{EX}$ after the avalanche is quenched. Obviously the SPAD detector is not able to detect another photon during the quenching cycle. The time between hit of the photon and the next possible detection is called dead time (t_D).

II. CHARACTERISTICS OF SPADs

Typical characteristics of SPADs are the dark count rate (DCR), the afterpulsing probability (APP) and the photon-detection probability (PDP).

A. Dark Count Rate and Afterpulsing Propability

The DCR defines the amount of counts not induced by a photon. These pulses can be generated either by uncorrelated interactions (e.g. thermally induced) between charge carriers and lattice defects of the semiconductor, so called charge traps, as described in [1], or correlated pulses which occur after a pulse. So called deep level traps are filled during the main avalanche of a pulse and after a statistical delay released carriers generate a secondary avalanche and therefore an afterpulse [2]. The lifetime of these carriers is in the nanosecond range and therefore the afterpulsing probability (APP) is directly correlated to the dead time of the QCs.

B. Photon Detection Probability

The quantum efficiency η describes the probability of generating an electron-hole pair out of a photon.

$$\eta = \frac{N_e}{N_p} = \frac{I_{ph}}{q \cdot \Phi} \tag{1}$$

where N_e describes the photo-generated electrons and I_{ph} the photocurrent. N_p is the number of incoming photons, Φ is the photon flux, and q the elementary charge.

For single-photon detectors, normally the PDP is used instead of the quantum efficiency. Analogous it is calculated by

$$PDP(\lambda) = \frac{N_d}{N_p(\lambda)} \tag{2}$$

with N_d is the number of detected photons and N_p the amount of incoming photons. Since the responsivity of the diode depends on the wavelength (λ), also the PDP depends on it. Other reasons for a PDP <1 are e.g. the photon is absorbed outside of the active area, it might be reflected on the detector surface, the photon transmits, it is absorbed in the active area, but is not detected due to recombination or other parasitic effects, or no self-sustaining avalanche happens.

978-1-7281-8494-4/20 $31.00 © 2020 IEEE

III. Integration of the SPAD

The SPAD is integrated in a high voltage 180 nm CMOS process.

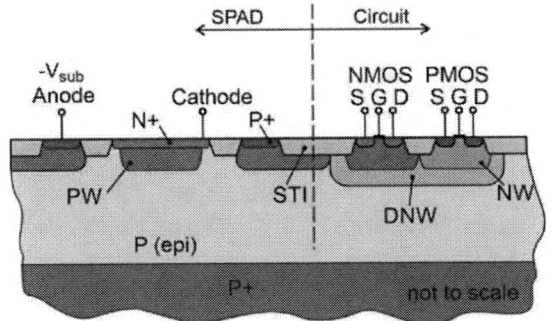

Fig. 1. Cross section of SPAD and concept of isolated transistors

The anode of the SPAD is directly connected to the substrate of the microchip, see Fig. 1, and therefore the QC has to be isolated from the substrate, due to a V_{BR} above the supply voltage of the circuit. The circuit is completely surrounded by deep N-wells (DNW) to ensure the isolation towards the substrate. The technology offers an isolation of 40 V between the substrate and the DNW.

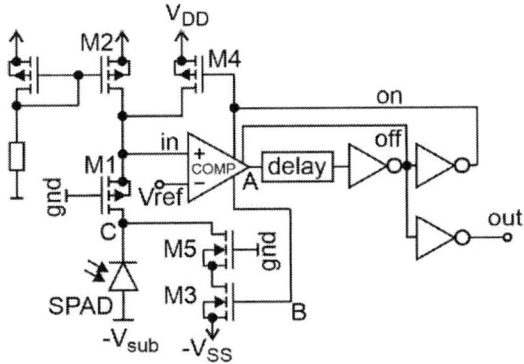

Fig. 2. Overview schematic of cascoded active quenching circuit

The SPAD cathode is connected to the QC, the multiplication zone is between the N+ and the P-well (PW), the absorption zone is in the PW and in the P-epi area. The P-epi of this process is higher doped compared to the opto-process reported in [3] and therefore the absorption zone is thinner, because the epi-layer is not fully depleted at the breakdown and excess bias voltage.

The diameter of the active area of the circular SPAD is 16 μm, while the complete SPAD occupies 1964 μm².

IV. Quenching Circuit

The QC is designed as cascoded active quencher. The schematic of the circuit is depicted in Fig. 2.

Node C is the cathode of the SPAD, which is close to V_{DD} when it is biased in Geiger mode, before a photon hits.

$$|V_{DD}| + |V_{sub}| = |V_{BR}| + |V_{EX}| \qquad (3)$$

The negative substrate voltage V_{sub} delivers the necessary bias voltage according to .

As soon as an event occurs, an avalanche is built up and the current through M2 leads to passive quenching (M4 is off),

which reduces the gate-source voltage of M1 and therefore the input voltage of the comparator (COMP) (and also V_C, see Fig. 4). As soon as it reaches $V_{ref} = 3.2V$, after 0.25 ns, he in Fig. 3 detects the event, V_A turns to "high" and turns off the pre-bias of V_B and it is pulled to "high" as well, but in the lower voltage regime, between ground and $-V_{SS}$, V_B is pre-biased to get a fast response time (see Fig. 5).

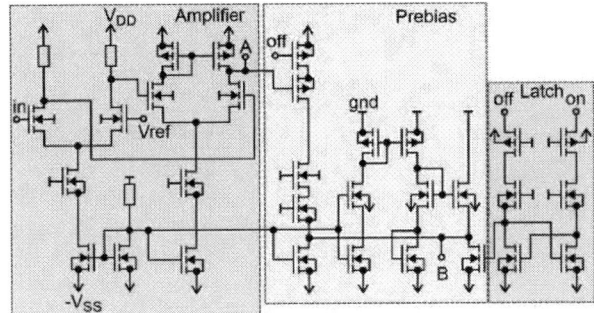

Fig. 3. Schematic of COMP

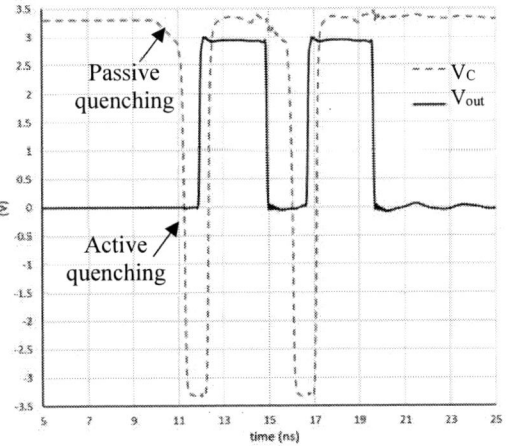

Fig. 4. Transient voltages (postlayout simulation), minimum dead time

Then, M3 (Fig. 2) is turned on, which decreases V_C rapidly, in 0.34 ns, to $-V_{SS}$, and therefore below V_{BR}, consequently, the avalanche is quenched. The overall quenching time betweenandis 1.34 ns. Fig. 5 depicts the transient voltages of the described nodes for the maximum dead time. In the meantime, the voltage at node A (V_A) starts the reset again. Via the delay block and the following inverters the dead time is variable. The voltage at node "on" (V_{on}) equals V_{DD}, except for the reset pulse close to the end of the deadtime, generated by the latch (Fig. 1) when the voltage of node B (V_B) is reset to $-V_{SS}$ again and V_C is pulled back towards V_{DD} by M4. The transistors M1 and M5, as well as the cascode transistors in the COMP, are necessary to keep all transistors in the save operating area. The COMP works similar to the active quenching circuit in-depth described in [3], which was realized in a 0.35 μm CMOS process with a thick low doped p-epi layer. All transistors are high-voltage transistors which enable a supply voltage of 3.3 V and due to the cascoded concept the total voltage swing is 6.6 V.

The circuit offers dead times from 4.5 ns (Fig. 4) to more than 30 ns (Fig. 5) but unfortunately the APP of the diode is rather high and the measurement results were best for the maximum dead time.

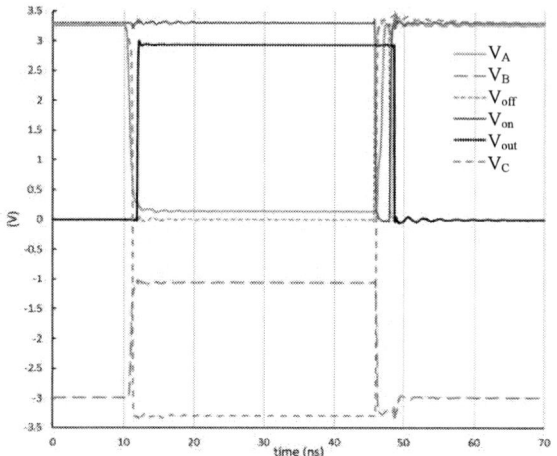

Fig. 5. Transient voltages (post layout simulation) for maximum dead time

Fig. 1 shows the microphotograph and a layout plot of the circuit. The active area of the quenching circuit inclusive SPAD is 185 μm × 150 μm.

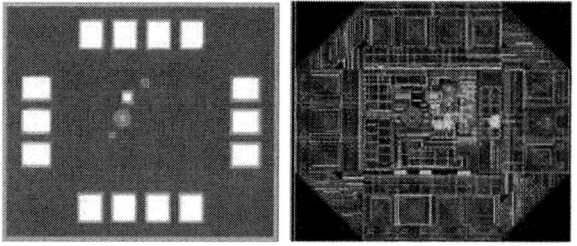

Fig. 6. Microphotograph and layout plot of the chip

V. MEASUREMENT SETUP

The device under test (DUT) was bonded onto a printed circuit board (PCB), which is mounted on a copper block for temperature stabilization. A temperature sensor placed within the copper block as well as a thermo-electric cooler are used to stabilize the DUT's temperature at 25°C. For the measurement, the DUT is placed in a dark box to remove all background light. Inside this dark box, also the detector of a Thorlabs power meter (PM200) and a motorized xyz stage are placed. For the PDP measurements, the laser light is guided to the DUT using a SM600 single-mode fiber. The mode-field diameter of this fiber is ~5 μm, which is considerably smaller than the active diameter of the DUT. By placing the end of the fiber only a few μm above the chip surface we can guarantee that almost all of the light from the fiber hits the active area of the SPAD .

The xyz stage is used to position the fiber over the center of the SPAD by scanning the chip surface with a constant optical power in the fiber and by finding the coordinates where most pulses are counted at the output of the quencher. Additionally, the xyz stage is used for a calibration measurement. It allows moving the optical fiber from the DUT to the detector of the Thorlabs optical power meter.

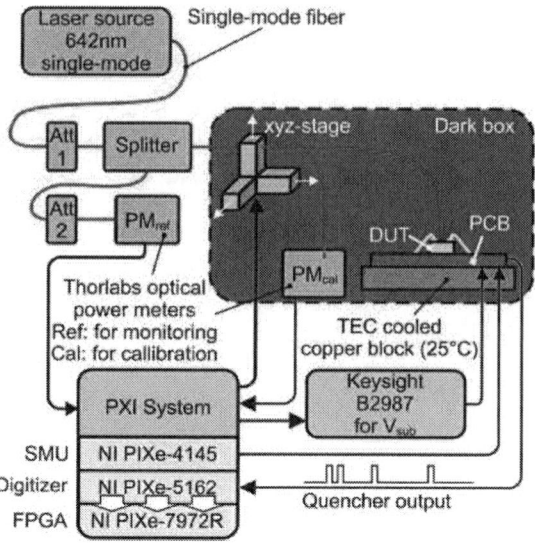

Fig. 7. Measurement set-up

For calibrating the optical power at the output of the single-mode fiber, two optical powermeters are used. The output of the 642 nm laser source is connected to a first attenuator (Att 1), which is followed by a fiber splitter. One output of the fiber splitter directly leads to the second optical power meter (Thorlabs PM100USB) that is placed outside the dark box. The second output is guided to a second attenuator (Att 2) that is connected to the fiber that guides the light to the DUT. During the calibration, the end of the fiber is placed over the detector of the optical power meter inside the dark box and the ratio between the optical power at the end of the fiber and the reference detector outside the box is determined. This ratio is larger than 10^5 in order to have sufficiently high optical power at the reference detector during the PDP measurement.

Using the reference detector and the fixed ratio calibration value allows monitoring the photon rate at the output of the fiber. After getting this calibration value, the attenuation of Att 1 is increased until the desired photon rate is reached.

A PXI system from National Instruments (NI) controls the whole measurement setup. The substrate voltage is applied by an electrometer from Keysight (B 2987), while the quencher is supplied by a source measurement unit (NI PXIe-4145) from NI. The output of the DUT is read-in by NI's digitizier (NI PXIe-5162) and is processed by directly streaming the raw data to a FlexRIO FPGA from NI (NI PXIe-7972R).

VI. MEASUREMENT RESULTS

For the characterization of the DUT, first dark measurements were done in order to measure dark count rate (DCR) and afterpulsing probability (APP). A pulse is counted as afterpulse count if it arrives within 200ns after the pulse before.

The breakdown voltage is defined as the voltage where the DCR is 1 count per second, which is in this case 13.2 V. Fig. 8 shows the DCR versus the excess bias voltage V_{ex}. It can be clearly seen that the dark count rate is rather high, compared

978-1-7281-8494-4/20 $31.00 © 2020 IEEE

to other technologies [3] where the DCR is significantly lower.

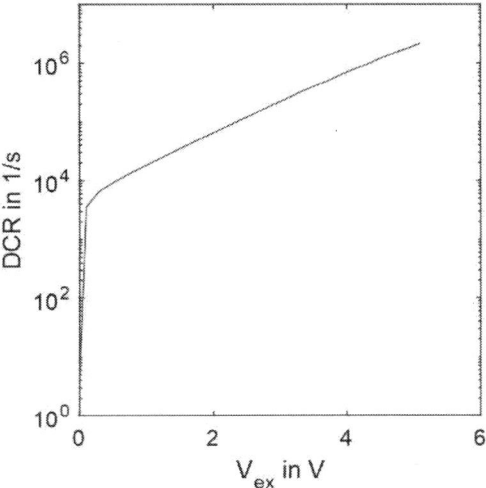

Fig. 8. DCR vs. excess bias

The reason for this might be the fact, that shallow trench isolation (STI) is used in the presented SPAD and therefore additional lattice defects and stress result in higher DCR, even for the very small active area.

The APP shown in Fig. 9 is again very high, for the same reasons as mentioned above.

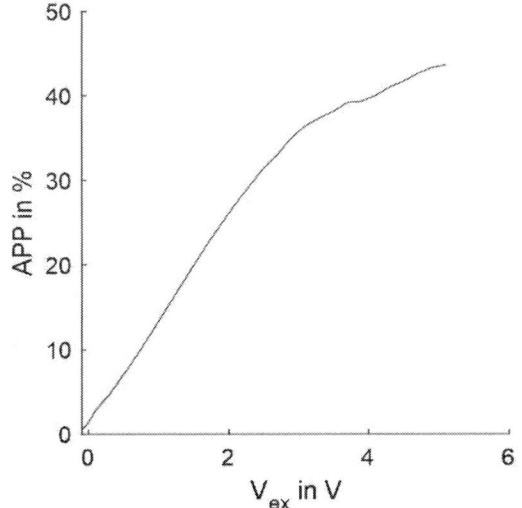

Fig. 9. APP vs. escess bias

For the PDP measurements, the fiber was placed a few micrometer above the SPAD's surface and the count rate was recorded depending on the substrate voltage. Additionally, the photon rate at the output of the fiber was monitored using the reference detector outside the dark box. The presented PDP was corrected for DCR, APP and saturation effects. In addition a moving average filter over 0.2V had to be applied

to smoothen the noisy PDP values for $V_{ex}>2.5V$ due to the high DCR and APP for these excess biases. There is a trend of overestimation of the APP, especially for large excess bias voltages, which results in an underestimation of the extracted PDP.

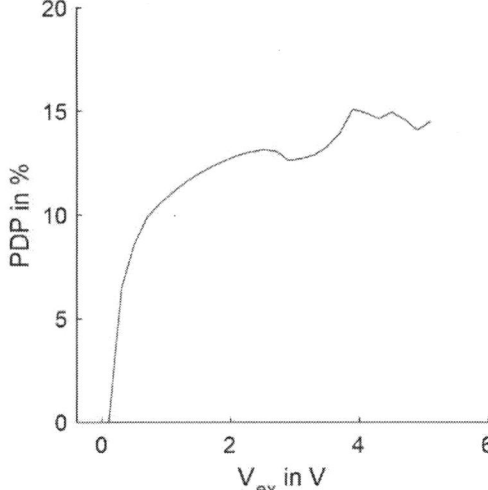

Fig. 10. PDP vs. escess bias at 642 nm.

VII. CONCLUSION

We present a fully integrated actively quenched SPAD in 180nm CMOS. While the current device still has a relatively high APP and DCR, we strongly believe that these parameters can be considerably improved by removing the STI close to the SPAD. Additionally, the DCR can be improved considerably be cooling the device. Reduced DCR and APP should also allow to further increase the PDP of the SPAD that already reaches 13 % at 642 nm at a relatively low excess bias voltage of 2.3 V and increases to 15.1 % for an excess bias voltage of 3.9V.

ACKNOWLEDGMENT

The authors acknowledge financial support from the Austrian Science Fund (FWF) under Grant No. P28335-N30. The authors would like to thank the colleagues at TU Wien for fruitful discussions and XFAB for chip fabrication.

REFERENCES

[1] S. Sze and K. Ng, Physics of Semiconductor Devices. Wiley, 3rd edition ed., November 2006.

[2] S. Cova, A. Lacaita, and G. Ripamonti, "Trapping phenomena in avalanche photodiodes on nanosecond scale," *IEEE Electron Device Letters,* vol. 12, pp. 685–687, Dec 1991.

[3] R. Enne, B. Steindl, M. Hofbauer and H. Zimmermann, "Fast Cascoded Quenching Circuit for Decreasing Afterpulsing Effects in 0.35-μm CMOS," in *IEEE Solid-State Circuits Letters,* vol. 1, no. 3, pp. 62-65, March 2018, doi: 10.1109/LSSC.2018.2827881.

[4] B. Steindl, M. Hofbauer, K. Schneider-Hornstein, P. Brandl and H. Zimmermann, "Single-Photon Avalanche Photodiode Based Fiber Optic Receiver for Up to 200 Mb/s," in IEEE Journal of Selected Topics in Quantum Electronics, vol. 24, no. 2, pp. 1-8, March-April 2018, Art no. 3801308, doi: 10.1109/JSTQE.2017.2764682.

38.5 Gb/s RoF Based Optical Receiver for 5G Mobile Remote Radio Head Applications

B. Mesgari, N. Vokić, B. Goll, B. Pichler, D. Milovančev, K. Schneider-Hornstein, H. Arthaber, and H. Zimmermann

Institute of Electrodynamics, Microwave and Circuit Engineering,
Vienna University of Technology, Gußhausstraße 25/E354, A-1040 Wien, Austria
Phone: +43-1-58801-354620, Fax: +43-1-58801-35499
E-mail:{baset.megari | nemanja.vokic| bernhard.goll | bernhard.e354.pichler | dinka.milovancev | kerstin.schneider-hornstein |
holger.arthaber |horst.zimmermann}@tuwien.ac.at

Abstract—A 38.5 Gb/s optical receiver has been fabricated in a 130 nm SiGe-BiCMOS technology for being employed in a Radio over Fiber transceiver for 5G remote radio head application. As a transimpedance topology, a common base amplifier is used to provide a low resistance input impedance to relax connecting a photodiode with relatively high capacitance. The transimpedance amplifier (TIA) is followed by the capacitive degeneration limiting amplifier to create a degree of freedom and relax the trade-off between the gain-bandwidth product and the linearity of the receiver. Analytical methods have been investigated to provide a low-noise and high gain design. Calibrated S-parameter measurement has been done which represents a 27-GHz 3-dB bandwidth with a differential gain of 49 dBΩ with a total power consumption of 160 mW for the entire receiver. The optical receiver has also been tested in the time domain by using a 35-GHz photodetector with a responsivity of 0.1 A/W and a capacitance of 90 fF. Eye diagram measurements have revealed that the optical receiver can provide a BER of 6.37×10^{-6}, 1.3×10^{-4}, and 1.7×10^{-3} for a data rate of 15, 25, and 35 Gb/s at an input current of 32µA respectively. The fabricated receiver occupies an area of 930 µm × 930 µm including bond pads.

Keywords— Radio over Fiber (RoF), 5G cellular communication, remote radio head (RRH), low noise optical receiver

I. INTRODUCTION

Radio over Fiber (RoF) technology has been significantly increased over the past few years in modern access networks such as 5G cellular communication, military radar, radio astronomy, and massive multiple-input multiple-output systems [1]-[3]. RoF technology is a combination of wireless communication and photonic devices to reduce system complexity and also enhance cost efficiency. In 5G wireless networks, a large number of pico-cells has to provide the utilization demands of the traditional base stations (BS) such as data traffic, broadband wireless coverage, and link throughput [2]. A wired connection, which is traditionally accomplished between the pico-cells network, might be impossible in most places in a city due to a large amount of cost and additional hardware. However, using a remote radio head (RRH) unit, which is an RoF transceiver, can provide a connecting link between pico-cells via a wireless or optical interface [3]. Fig. 1 exhibits the block diagram of a time division multiple access (TDMA) RoF transceiver which contains two main paths namely uplink and downlink. In the transmit time window (uplink) the received light signal from

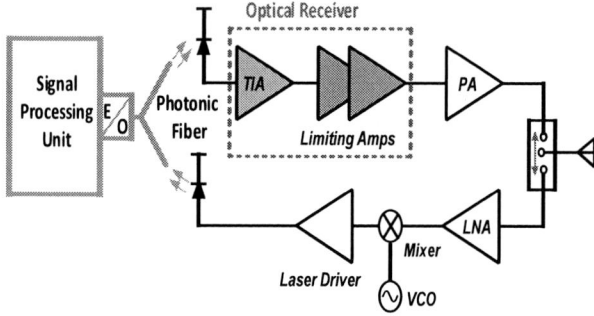

Fig. 1. Block diagram of a time division multiple access (TDMA) RoF transceiver.

BS is converted to electrical current by using a photodetector (PD) which is located at the input of the optical receiver (Fig. 1) [2]. Since the optical fiber between BS and RRH might introduce signal quality degradation such as path loss, the optical receiver with minimum possible noise performance is quite vital [1]. On the other side, the output voltage of the optical receiver should be large enough to drive the power amplifier (PA) at its optimum input voltage such that the maximum deliverable efficiency can be obtained by PA while the transmit mask specification has still remained acceptable. It can be concluded that the design of the optical receiver in this system is relatively challenging because it should present the ultra-low noise performance, large bandwidth, and significantly high-linear gain simultaneously.

Fig. 2 demonstrates the block diagram of the optical receiver topology used in this paper which is called pseudo-differential [4]. The incoming received light from BS is converted to a photocurrent by a PD and then the mentioned generated current should be amplified and converted to a voltage by using a transimpedance amplifier (TIA) [5],[6]. In this receiver topology a dummy TIA is utilized to generate a differential signaling, to suppress the common-mode noise, and enhance the gain with a factor of two approximately. In addition, by utilizing a differential topology, the receiver's tolerance against process and temperature variations will be increased [4], [6]. The TIA is followed by two limiting amplifiers (LA) and a 50-ohm output buffer in order to present sufficient linear gain for the input of PA. As it has been highlighted in Fig. 2, since the optical receiver is not completely differential, an offset compensation loop is used which can diminish the offset voltage due to process and temperature variations [6].

978-1-7281-8494-4/20 $31.00 © 2020 IEEE

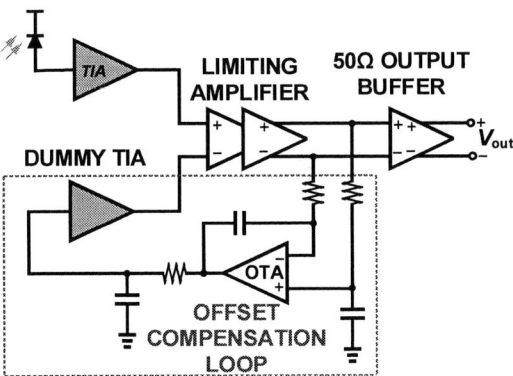

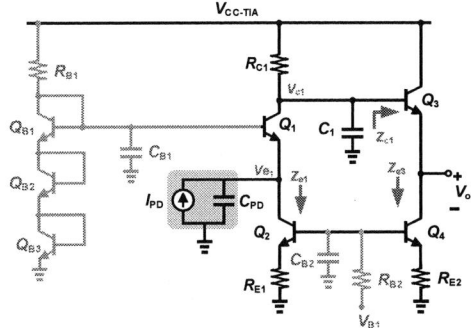

Fig. 3. Circuit diagram of the common-base TIA including bias circuitry.

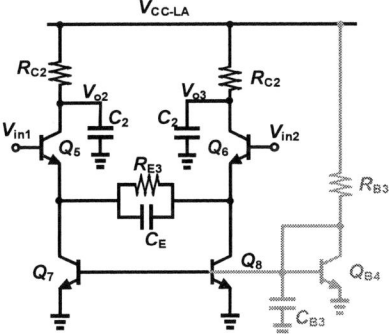

Fig. 2. Block diagram of the optical receiver system with offset compensation loop.

The paper is organized as follows. The analysis of the optical receiver is presented in section II. Section III is devoted to the results of the measurements. In this section, the time domain and frequency domain results by means of the eye diagram and scattering parameter, respectively, will be exhibited. Finally, conclusions are provided in section IV.

II. THE RECEIVER CIRCUIT ANALYSIS

A. Frequency response of the entire receiver

This section deals with the design of the SiGe-BiCMOS optical receiver circuit for 38.5 Gb/s. Some important theoretical aspects of the receiver circuit will be presented such as frequency response and noise behavior of the TIA. Fig. 2 represents the block diagram of the optical receiver system. The TIA is the first block in the receiver chain. Thus, wide bandwidth, high transimpedance gain, low noise, and low input-impedance of the TIA define the performance of the optical receiver. Due to the large diameter of plastic optical fibers and the need for ease of connecting fibers to the PDs as well as large light-sensitive areas of optical free-space receivers, the PDs are usually designed with large diameters. This results in a large input node capacitance of the TIA, in limited bandwidth and in increased input-referred noise. As a result, the input-impedance of a TIA should be as low as possible to avoid creating an unwanted dominant pole at the input port of an optical receiver. Although, many input impedance reduction techniques have been investigated based on the negative feedback and negative resistance topology, in high-frequency regime instability might happen due to a large number of parasitic paths with positive phase shift [7]. Hence the simpler the topology to choose as a circuit, the more stable it can be. Consequently, in this design, a common-base (CB) TIA and a conventional fully differential LA in combination with bandwidth extending parallel RC filter (capacitive degeneration) have been employed for implementing the 38.5 Gb/s optical receiver as shown in details in Fig. 3 and Fig. 4, respectively. It should be mentioned, large gain at the output of the limiters might lead to a saturated signal at the output of buffer, hence an offset cancelation loop (see Fig. 2) is essential in this topology.

Considering the desired power at the input of the PA shown in Fig.1, and generated current from PD, the transimpedance gain Z_t of the entire optical receiver can be calculated using equation (1). Z_{t-TIA}, Z_{t-LA} and Z_{t-BUF} express the TIA, LA and buffer gain, respectively. Based on the circuit diagram shown in Fig. 3, Z_{t-TIA} can be obtained using (2),

Fig. 4. Circuit diagram of the single stage capacitive degeneration limiting amplifier.

where g_{m1}, g_{m3}, are the transconductances of Q_1, Q_3, respectively. C_1 and C_{in} represent the total parasitic capacitances of node V_{c1} and V_{e1} respectively considering the Miller effects of base-collector junction. Before obtaining the 3-dB bandwidth of the Z_{t-TIA}, seen impedance of Z_{e1}, Z_{c1}, and Z_{e3}, associated to the V_{e1}, V_{c1}, and V_{c3} nodes, respectively, are assumed relatively high which can lead to a simplification of (2) investigating an intuitive insight of circuit behavior. For sake of simplicity, also, r_o, r_π ,C_μ for all transistors have not been considered. In this equation, Q and ω_n are quality factor and natural frequency of the Z_{t-TIA} transfer function.

$$Z_t = Z_{t-TIA} \times Z^2_{t-LA} \times \underbrace{Z_{t-BUF}}_{=1} \tag{1}$$

$$\frac{V_{o1}}{I_{PD}} = g_{m1}\left(Z_{e1} \| \frac{1}{(g_{m1}+SC_{in})}\right)\times\left(R_{C1} \| Z_{c1} \| \frac{1}{SC_1}\right)\times\underbrace{\frac{g_{m3}Z_{e3}}{g_{m3}Z_{e3}+1}}_{=1}$$

$$\simeq \frac{R_{C1}}{1+\underbrace{\left(R_{C1}C_1+\frac{C_{in}}{g_{m1}}\right)}_{\frac{1}{Q\omega_n}}S+\underbrace{\left(R_{C1}C_1\frac{C_{in}}{g_{m1}}\right)}_{\frac{1}{\omega_n^2}}S^2}$$

$$\tag{2}$$

The 3-dB bandwidth of the TIA can be calculated as (3).

$$BW_{3-dB} = \frac{\omega_n}{\sqrt{2}Q}\sqrt{-\left(1-2Q^2\right)+\sqrt{\left(1-2Q^2\right)^2+4Q^4}} \tag{3}$$

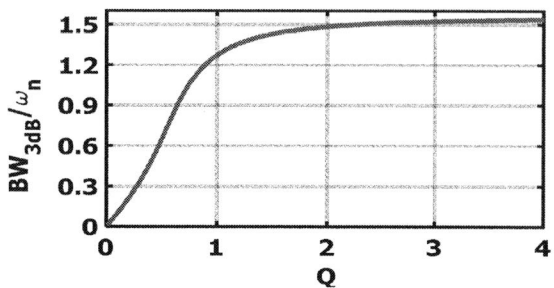

Fig. 5. Ratio of BW_{3dB}/ω_n as a function of Q.

As can be seen in Fig. 5, low values of Q lead to extreme bandwidth reduction while high values of Q have no significant effect on bandwidth improvement. Besides, large Q values lead to overshoot and ringing in the time-domain response of the circuit. E.g. the overshoot approaches 44% for a Q factor of 2. Therefore, it is best to choose Q between 0.7 and 1 to achieve the required bandwidth and keep a proper time-domain response [5]-[8].

Considering Fig. 4, the LA's voltage gain is expressed in (4).

$$\frac{V_{o3}-V_{o2}}{V_{in1}-V_{in2}} \simeq \frac{2g_{m5}R_{C2}\left(1+R_{E3}C_E S\right)}{\left(2+g_{m5}R_{E3}\right)\left(1+R_{C2}C_2 S\right)\left(1+\dfrac{2R_{E3}C_E S}{2+g_{m5}R_{E3}}\right)} \quad (4)$$

The parallel combination of R_{E3} with C_E shown in Fig. 4, can provide two main advantages, first compensating the gain roll-off resulting from the pole at the output node which means based on equation (4), if ($R_{E3}C_E=R_{C2}C_2$), the dominant pole can be defined at the emitter of the Q_5 and Q_6. Secondly, since the linear gain at the input of the PA is desired, therefore R_{E3} can increase the linearity of the LAs. As a result, the gain-bandwidth product trade-off of this topology which is ($g_{m5}R_{C2}/R_{E3}C_E$) has a degree of freedom which would not be achieved without using R_{E3} and C_E.

B. Noise Analysis of the TIA

The TIA is the first block in the receiver chain. Thus, the low noise performance of the entire receiver mainly depends on the noise behavior of the TIA [5]-[6]. Therefore, the meaningful portion of noise behavior of the entire receiver can be estimated just by analyzing the TIA. The equivalent small-signal model of the TIA circuit is depicted in Fig. 6, including the noise sources of Q_1, Q_2. Only the thermal noise of the resistors (R_{C2} and r_{b1}) and the collector shot noise are considered, while flicker noise and base shot noise are neglected for the sake of simplicity. The PSD of a resistor and a transistor noise current, is expressed as (5) and (6), respectively. Where K is the Boltzmann's constant, T expresses the absolute temperature, r_{b1} is the base resistance and R and g_m represent the resistance and transconductance of the resistor and transistor, respectively.

$$\frac{\overline{I^2_{nR}}}{\Delta f} = \frac{4KT}{R} \quad (5)$$

$$\frac{\overline{I^2_{nQ}}}{\Delta f} = 2qI_c = 2KTg_m \quad (6)$$

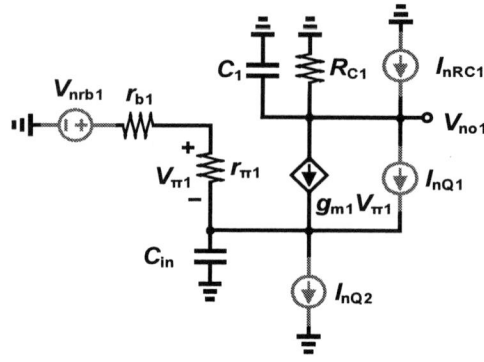

Fig. 6. The equivalent small-signal model of the CB TIA including the noise sources for calculating output noise.

According to Fig. 6, and using (5), (6) the output referred noise voltage can be obtained as follows in (7).

$$\overline{V_{n,out}^2} = \left(\overline{I^2_{n,RC_1}}\times|T_1|^2 + \overline{I^2_{n,Q_1}}\times|T_2|^2 + \overline{I^2_{n,Q_2}}\times|T_3|^2 + \overline{V^2_{n,rb_1}}\times|T_4|^2\right)$$

$$T_1 = \frac{R_{C1}}{\left(1+R_{C1}C_1 S\right)}$$

$$T_2 = \frac{T_4}{g_{m1}} = \frac{R_{C1}C_{in}S}{g_{m1}\left(1+\dfrac{C_{in}}{g_{m1}}S\right)\left(1+R_C C_1 S\right)}$$

$$T_3 = \frac{R_C}{\left(1+\dfrac{C_{in}}{g_{m1}}S\right)\left(1+R_C C_1 S\right)}$$

$$(7)$$

The integrated root mean square (rms) value of the input referred noise current $I_{inir\text{-}rms}$ of the common-base TIA can be obtained by integrating the output voltage of equation (7) from 0 to ∞ which is summarized in equation (8).

$$I_{inir-rms} = \frac{\sqrt{\displaystyle\int_0^\infty \overline{V_{n,out}^2}(f)df}}{R_{C1}}$$

$$= \sqrt{\frac{KT}{C_1 R_C^2}\left(\frac{2+4\times g_{m1}r_{b1}}{\left(\dfrac{1}{g_{m1}R_{C1}}+\dfrac{C_1}{C_{in}}\right)} + \frac{g_{m1}R_{C1}}{\left(\dfrac{C_{in}}{C_1 g_{m1}R_{C1}}+1\right)}+1\right)} \quad (8)$$

III. EXPERIMENTAL RESULTS

The optical receiver which is illustrated in Fig. 2 has been fabricated in 0.13 μm SiGe-BiCMOS technology with seven metal layers. Die micrograph of the fabricated receiver occupies 930 μm × 930 μm including bond pads. The circuit parameters such as the size of transistors, capacitors, and resistors as well as the biasing voltages are optimized using SPECTRE simulations and equations (1)-(8). The chip has been mounted on a test board with wire-bonds for all pads including 35-GHz pin PD with a responsivity of 0.1 A/W and a capacitance of 90 fF (see Fig. 7(a)).

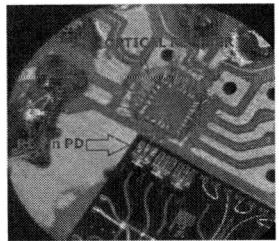

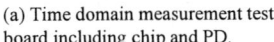

(a) Time domain measurement test board including chip and PD.

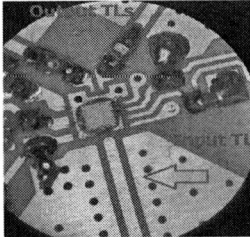

(b) S-parameter measurement test board depicting TLs.

Fig. 7. Measurement test boards.

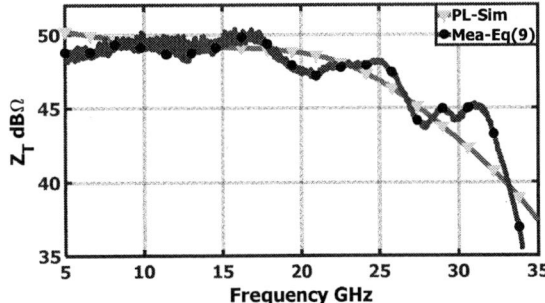

Fig. 9. Extracted Z_T using equation (9) based on S-parameter. For comparison the post-layout simulation result is also included.

The optical receiver consumes 160 mW. For characterizing the optical receiver two different methods have been employed namely S-parameter and eye diagram measurements. In the S-parameter measurements, the input and output pads, have been bonded to 50Ω coplanar waveguide transmission lines (TLs) on the test board using minimum-length bonding wires (Fig.7b). The input and output ports of the test board were connected to a 67 GHz PNA-X microwave network analyzer via 50 GHz 2.4 mm coaxial connectors. In order to enhance the accuracy of the measurement, a TRL calibration board was designed for de-embedding the degradation caused by connectors, TLs, and test board parasitics. The calibrated S-parameters are illustrated in Fig. 8. When the optical receiver is directly connected to 50Ω loads, the transimpedance gain Z_T can be expressed as (9) [9]. For achieving the differential gain, 6 dB should be added to the S_{21} results.

$$Z_T = \frac{Z_{21}}{\left(1 + \dfrac{Z_{22}}{Z_0}\right)} = Z_0 \frac{S_{21}}{1 - S_{11}} \qquad (9)$$

Where Z_0 equals to 50Ω and Z_{21} and Z_{22} are the Z-parameters of the optical receiver. Fig. 9 shows the Z_T of the entire optical receiver using equation (9). As illustrated in Fig.9, a differential gain of 49 dBΩ is achieved with the 3-dB bandwidth of 27 GHz which means the optical receiver can provide a 38.5 Gb/s data rate for a non-return to zero bit pattern.

Time-domain characterization has also been done using a 40 Gb/s pseudorandom bit sequence (PRBS) pattern generator as an input source and a DSA-8200 digital serial analyzer for measuring the eye diagram at the output of the receiver.

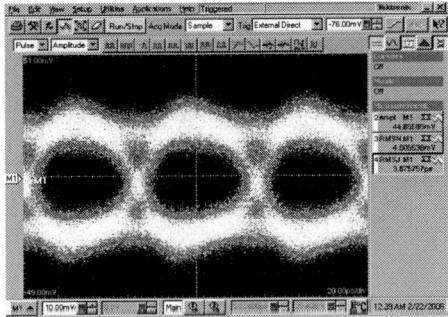

(a) 15 Gb/s

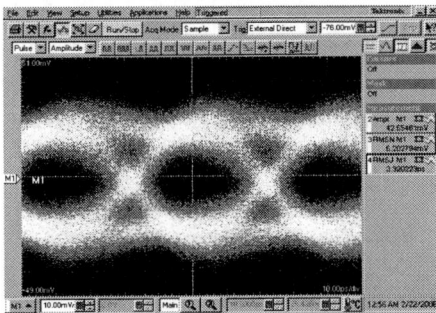

(b) 25 Gb/s

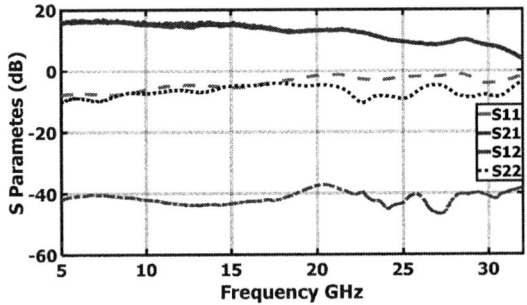

Fig. 8. Calibrated S-parameter measurement result of the optical receiver.

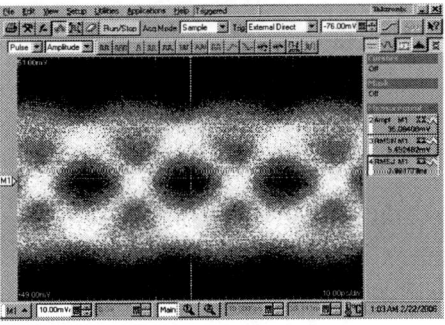

(c) 35 Gb/s

Fig. 10. Eye diagram measurements for three different bit rates (15, 20 and 35 Gb/s) based on PRBS 2^{15}-1 and 32μA input current using a 35GHz PD which is shown in Fig.7a.

Three different bit-rates (15, 20, and 35 Gb/s) have been tested with the input current of 32μA and the obtained eye

978-1-7281-8494-4/20 $31.00 © 2020 IEEE   69

diagrams are shown in Fig. 10. Based on [6] the bit error rate (BER) has been calculated considering V_{p-p} and σ, which were obtained based on the above eye diagrams plots directly with the DSA-8200. BER can be calculated by using (10) while assuming the same noise for both ones and zeros. The below equation is accurate when $Q_F \geq 2$.

$$BER \simeq \frac{1}{\sqrt{2\pi}} e^{\left(\frac{-Q_F^2}{2}\right)} \left(1 - \frac{1}{Q_F^2}\right) \qquad (10)$$

$$Q_F = \frac{V_{p-p}}{2\sigma}$$

The calculated BERs based on equation (10) have been summarized in Table 1. As it is illustrated in Table 1, the best BER has been archived for 15 Gb/s, which is 6.37×10^{-6} at the input current of 32μA.

TABLE I.   BER MEASUREMENT SUMMARY

Gb/s	V_{p-p} (mV-rms)	σ (mV-rms)	Q_F	BER
15	44.65	4.8	4.65	6.37×10^{-6}
25	42.55	5.2	4.09	1.3×10^{-4}
35	35.09	5.42	3.31	1.7×10^{-3}

IV. CONCLUSION

In this paper, a wideband low-pass optical receiver has been fabricated in a 130 nm SiGe-BiCMOS technology to be employed in a Radio over Fiber transceiver for 5G remote radio head application. In this receiver topology, a dummy TIA is utilized to generate a differential signaling, to suppress the common-mode noise, and to enhance the gain with a factor of two approximately. A simple CB topology has been used to provide a stable and low impedance path at the input port of the receiver. A conventional fully differential LA in combination with bandwidth extending parallel RC filter (capacitive degeneration) has been employed to amplify the converted signal at the output of TIA to the desired level suitable for PA. For characterizing the optical receiver, S-parameter and eye diagram measurements have been done. Calibrated S-parameter measurements show a differential gain of 49 dBΩ in a 3-dB bandwidth of 27-GHz. The receiver consumes a total power of 160 mW while occupying 930 μm × 930 μm including bond pads. BERs, which allow error correction, using eye diagram measurements for a 35-GHz pin PD with a responsivity of 0.1 A/W and a capacitance of 90 fF are achieved for different data rate of 15, 25, and 35 Gb/s at an input photocurrent of 32μA.

ACKNOWLEDGMENT

The authors acknowledge financial funding from the Austrian BMVIT via FFG in the project TRITON. They also thank Franz Dielacher from Infineon IFAT for access to the design environment and for chip fabrication.

REFERENCES

[1] Minkyu Sung, *et al.*, "RoF-Based Radio Access Network for 5G Mobile Communication Systems in 28 GHz Millimeter-Wave," *IEEE JLT, vol. 38, no. 2, pp.409–420, Jan. 2020.*

[2] Mahmood Noweir, et al., "Digitally Linearized Radio-Over Fiber Transmitter Architecture for Cloud Radio Access Network's Downlink,,". *IEEE Trans. Microw. Theory Tech., vol. 66, no. 7, pp. 3564–3574, July 2018.*

[3] Yipeng Wang, *et al.*, "A 60-GHz 4-Gb/s Fully Integrated NRZ-to-QPSK Fiber-Wireless Modulator," *IEEE Trans. Circuits Syst. 1, vol. 64, no. 3, pp. 653–663, March. 2017.*

[4] B. Mesgari, *H. Mahmoudi, and H. Zimmermann*, "A Single-to-Differential Transimpedance Amplifier for Low-Noise and High-Speed Optical Receivers" 2019 Austrochip Workshop on Microelectronics (Austrochip), pp.76-80.

[5] B. Razavi, Design of Analog CMOS Integrated Circuits. New York: McGraw-Hill, 2001.

[6] H. Zimmermann, Silicon Optoelectronic Integrated Circuits. Springer, Berlin, Heidelberg, 2004.

[7] *B. Abdollahi, B. Mesgari, S. Saeedi, and A. Nabavi, "Stability analysis and compensation technique for low-voltage regulated cascode transimpedance amplifier," Microelectronics J., vol. 71, November 2017, pp. 37–46.*

[8] *B. Abdollahi, P. Akbari, B. Mesgari, and S. Saeedi, "A low voltage low noise transimpedance amplifier for high-data-rate optical recievers," in 23rd Iranian Conference on Electrical Engineering, 2015, pp. 1187–1192.*

[9] Charles Q. Wu, *et al.*, "40-GHz Transimpedance Amplifier With Differential Outputs Using InP–InGaAs Heterojunction Bipolar Transistors" *IEEE J. Solid-State Circuits, vol. 38, no. 9, pp. 1518–1523, Sep. 2003.*

Comparison of Bandwidth Extension Methods for Doherty Power Amplifiers for 5G

Sina Mortezazadeh Mahani [*], David Seebacher [†], Matteo Bassi [†], Johannes Sturm [*]

[*]Silicon Austria Labs GmbH, Villach, Austria

[†] Infineon Technologies, Austria AG, Villach, Austria

Email: {*Sina.Mahani, J.Sturm*}@silicon-austria.com,{*David.Seebacher, Matteo.Bassi*}@infineon.com

Abstract—**This paper presents a systematic analysis overview based on theoretical derivations and simulation results for different versions of Doherty Power Amplifier (DPA). The Bandwidth (BW) for average efficiency and maximum output power level is evaluated and compared. The comparison is between the conventional current-combined DPA and voltage-combined DPA with applying the BW enhancement method for broadband 5G wireless transmitter application. The evaluations are done at 27 GHz center frequency. It is concluded that the voltage-combined DPA modified with efficiency in a generic way has one bandwidth-limiting factor that is due to the practical restrictions of transformer implementation.**

Index Terms—**Doherty PA, Doherty output network, broadband PA, efficiency enhancement, 5G PA**

I. INTRODUCTION

To meet the ever-increasing demand of data rate within limited spectrum resources for 5G wireless communications, high-order modulations are employed. However, the resulting high peak-to-average power ratio (PAPR) waveforms pose stringent requirements on PBO efficiency [1]. Among various PBO enhancement techniques, the Doherty PA is a promising candidate that can support a large modulation bandwidth and requires a low baseband digital processing overhead [2]. Since its introduction in 1936, DPA has been extensively explored and become one of the most widely used PA architectures in existing cellular base stations [3]. Recent research shows that DPA has the capability of operating at mm-wave frequencies [4], conventional DPA has several factors that are contributing to its BW limitation, therefore BW extension is an important aspect in modern DPA designs, and it has received increasing attention in recent researches. This subject is even more critical for 5G wideband applications. There are many articles and literature reviews on this topic during the past several years [5]-[6]. However a systematic comparison on broadband DPA designs has not been studied that is an essential subject for 5G wireless transmitters. Therefore a comprehensive review on all modifications of DPA with different Doherty output network regarding broadband applications could be useful. We provide a new insight through a theoretical derivation and simulation for symmetric DPA structures with respect to the variation of presented load to main amplifier in DPA structure, that is the main limiting factor for bandwidth of average efficiency and maximum output power level. Different topologies of Doherty

output networks are evaluated under the same condition for 1 V DC power supply at 27 GHz center frequency.

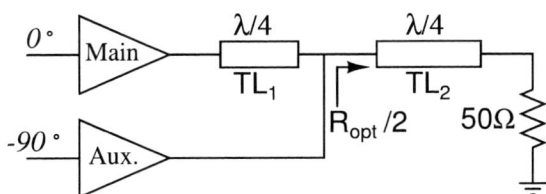

Fig. 1: Block diagram of conventional Doherty PA.

II. CONVENTIONAL DOHERTY OUTPUT NETWORK

In conventional DPA architecture (Figure 1) both main and auxiliary amplifiers see their optimum load resistance to deliver maximum power and efficiency. With respect to the output power level, the combiner node impedance (R_{comb}) is $R_{opt}/2$ for low output power. The auxiliary amplifier turns on at desired back-off efficiency point and $R_{comb} \rightarrow R_{opt}$, therefore the DPA delivers its maximum power to the load. The main bandwidth-limiting components are TL_1 and TL_2. This can be shown by exploring R_{comb} when the Doherty output network is operating [3]. Broadband operation of Doherty output network is distinguished by flatness of output power and average efficiency over frequency. The variation of the maximum deliverable output power and the back-off efficiency versus frequency are due to the variation of the load that is presented to the main amplifier. Therefore, the lower variation of the load presented to amplifier versus frequency delivers a wider bandwidth for DPA. Considering the ideal $\lambda/4$ transmission lines, the conventional Doherty network has zero power variation versus frequency for a certain output power level. This output power level will be referred as Optimum power level ($P_{opt,BW}$) for which Doherty network can work with optimum flatness over bandwidth. The R_{opt} for main amplifier delivering a specified output power level is given by Equation (1).

$$R_{opt} = \frac{V_{dc}^2}{2 \cdot (10^{\frac{P_{Max,main}(dBm) - 3}{10}})} \tag{1}$$

978-1-7281-8494-4/20 $31.00 © 2020 IEEE

In Equation (1), V_{dc} is DC supply voltage and is limited by the technology. $P_{Max,main}$ is the maximum deliverable output power level in dBm for main amplifier. The maximum output power delivered to the load of the Doherty is 3 dB more than $P_{Max,main}$. In conventional DPA, when $R_{opt} = 100\,\Omega$ the maximum output power variation is ideally zero. This output power level for DPA is a moderately low to what applications demand. The load variation versus frequency in

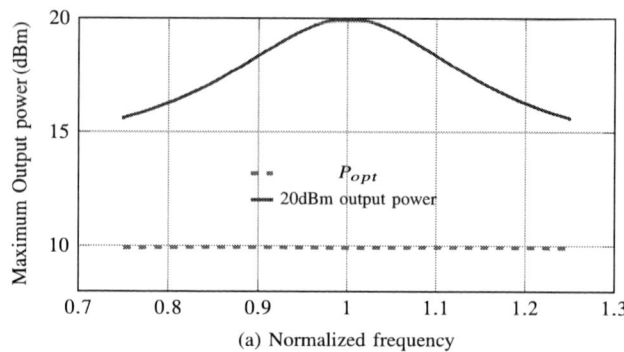

(a) Normalized frequency

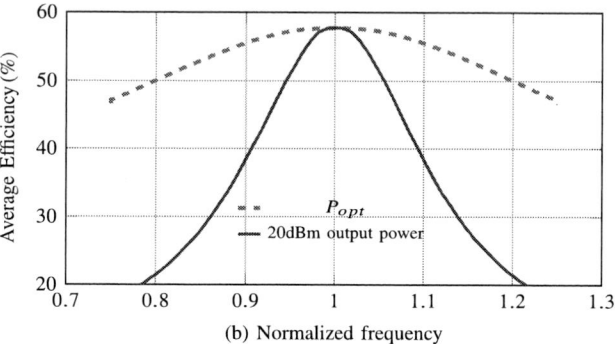

(b) Normalized frequency

Fig. 4: Evaluation of the performances over frequency bandwidth (a) maximum output power level, (b) average efficiency versus normalized frequency.

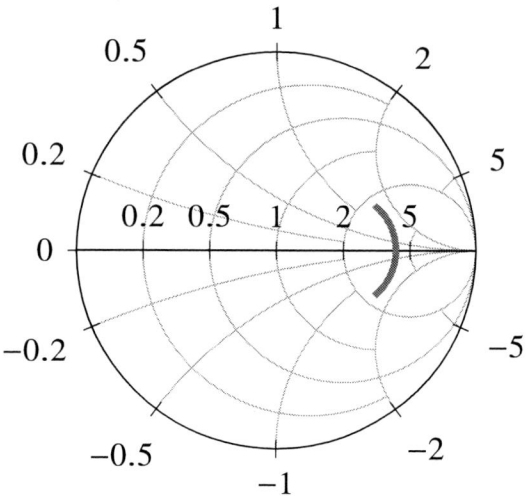

Fig. 2: The variation of the presented load to the main amplifier over frequency at back-off efficiency point.

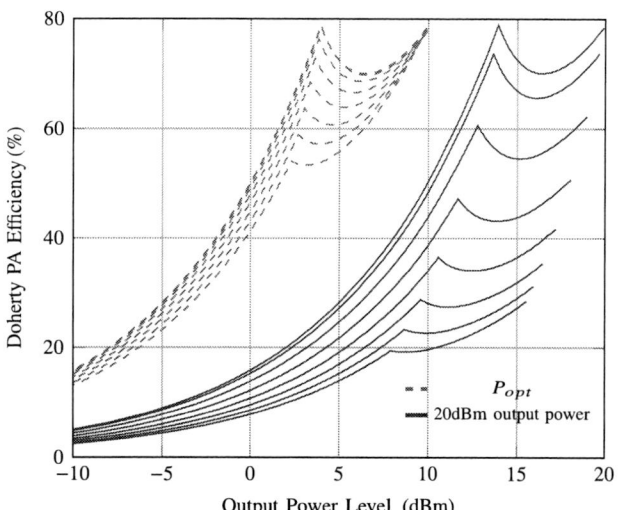

Fig. 3. Efficiency versus output power level.

Figure 2 at back-off efficiency is the main bandwidth limiting factor for conventional DPA. Figure 3 shows that for higher output power level the bandwidth of back-off efficiency and maximum output power level drops drastically.

As a criteria, the average efficiency and maximum output power are evaluated for bandwidth enhancement approaches and other topology of the Doherty output network for compar-

ison with the bandwidth of the conventional DPA. The average efficiency is evaluated based on Rayleigh pdf with 9 dB PAPR [7].

III. EVALUATION OF BANDWIDTH ENHANCEMENT METHODS FOR CURRENT-COMBINED DPA

In [8], a $\lambda/4$ short-circuited stub connected to the combiner node improved the BW of backoff efficiency, as shown in Figure 5. This stub modifies the impedance presented to TL_1 and compensate the impedance drop experienced by the main amplifier. A parallel LC connected to the combiner node as it is shown in Figure 5 has similar impact. This bandwidth enhancement approach reduces the variation of the back-off efficiency point or average efficiency over frequency. Figure 6 shows the amount of bandwidth enhancement on the performances for the proposed topology in [8]. It is observed that BW degradation for maximum output power is negligible.

Another modified DPA architecture for broadband application is inverted DPA with two $\lambda/4$ transmission lines for auxiliary amplifier, as shown in Figure 7. Inverted topology could improve the level of P_{opt}, since the impedance transformation ratio of TL_1 in Figure 7 is reduced. Figure 8 displays a considerable reduction of variation of the load presented to main amplifier at back-off efficiency compared to conventional DPA. A substantial improvement for BW of evaluated performances for inverted DPA in comparison with conventional DPA is observed in Figure 9. It can be seen that P_{opt} has increased roughly 2 dB by using inverted topology. The BW enhancement method is realized with modifying inverted DPA

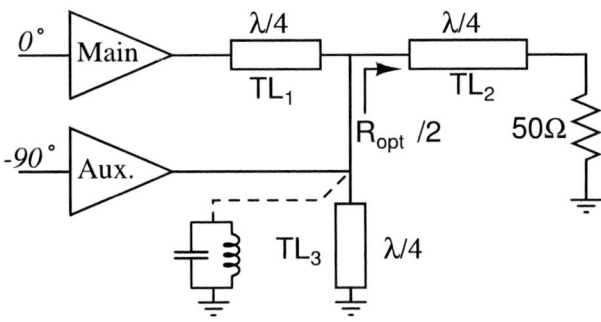

Fig. 5: modification for bandwidth extension of the back-off efficiency.

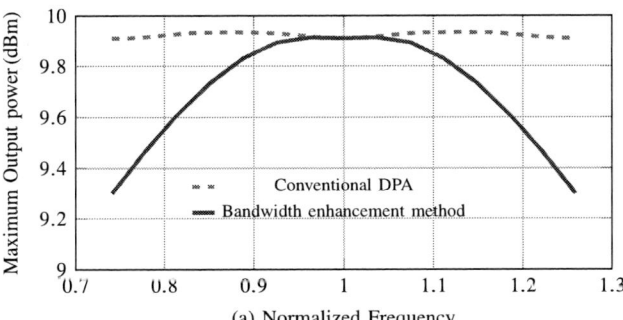

(a) Normalized Frequency

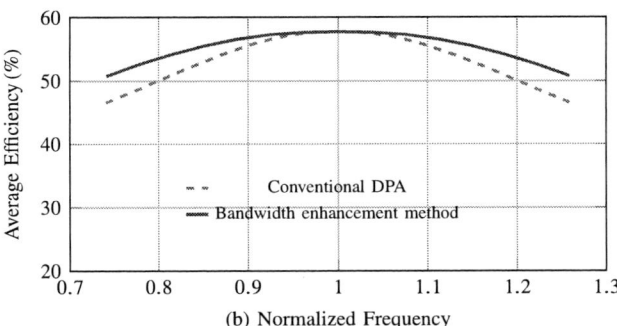

(b) Normalized Frequency

Fig. 6: Evaluation of the performances for proposed topology in [8] (a) Maximum output power level (b) Average efficiency versus normalized frequency.

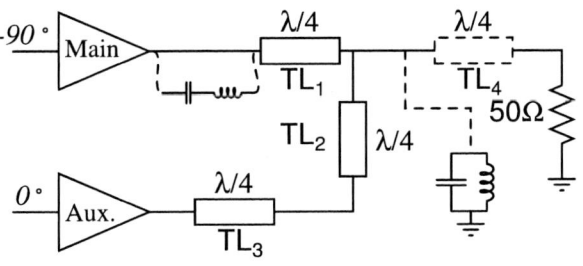

Fig. 7: Inverted DPA block diagram with possible modifications (dashed) for bandwidth improvement.

with a parallel LC resonator connected to the combiner node or a series LC resonator in main amplifier branch in presence of TL_4 as shown in Figure 7. This modification yield to improve

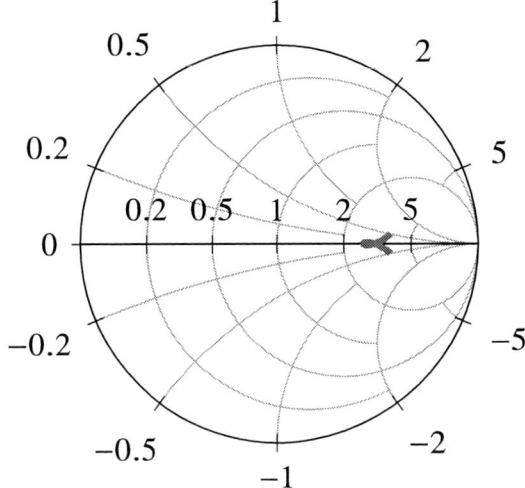

Fig. 8: The variation of the presented load to the main amplifier over frequency at back-off efficiency point.

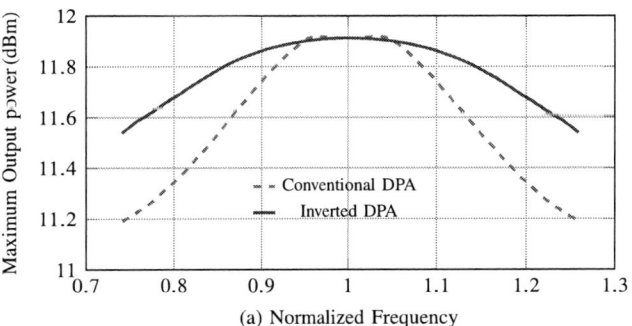

(a) Normalized Frequency

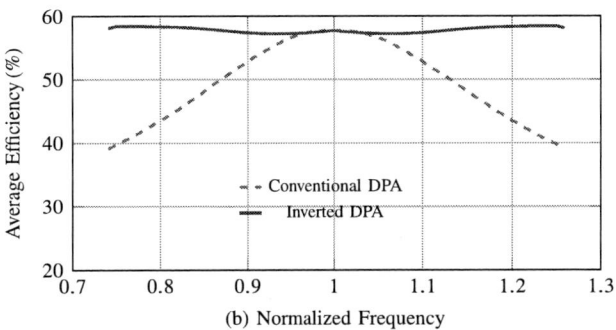

(b) Normalized Frequency

Fig. 9: (a) Maximum output power level (b) Average efficiency versus normalized frequency.

BW for higher output power level compared to inverted DPA.

IV. EVALUATION OF BANDWIDTH ENHANCEMENT METHODS FOR VOLTAGE-COMBINED DPA

Figure 10 shows a transformer-based series output combiner used in [9]. In comparison with conventional current combined DPA, voltage-combined DPA has low impedance transformation ratio from $50\,\Omega$ to $2R_{opt}$ [10], which results in larger BW [11]. The reduction of impedance transformation ratio in Voltage-combined topology (Figure 10 (a)) has a

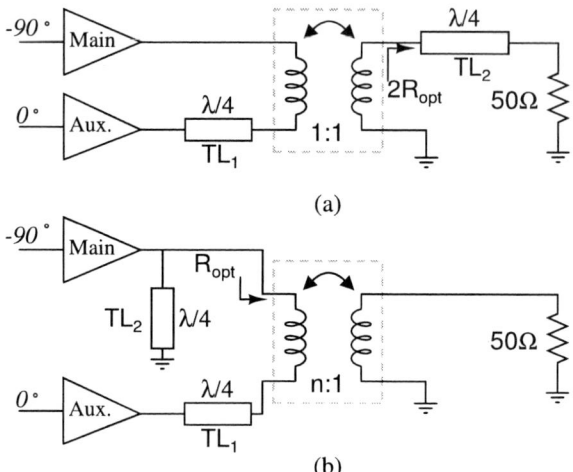

(a)

(b)

Fig. 10: (a) Voltage-combined DPA (b) with bandwidth enhancement method.

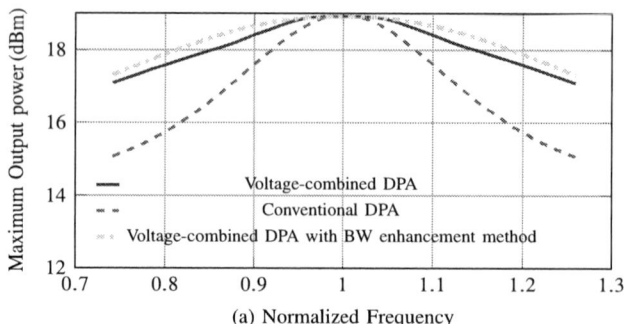

(a) Normalized Frequency

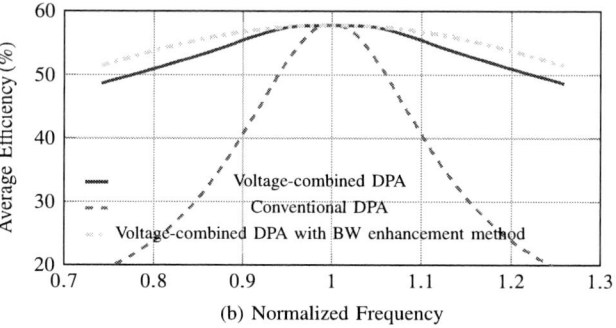

(b) Normalized Frequency

Fig. 11: (a) Maximum output power level (b) Average efficiency versus normalized frequency.

considerable effect on the BW of DPA. The desired performances over frequency are shown in Figure 11. The bandwidth enhancement method as shown in Figure 10 (b) is a promising way to improve the voltage-combined DPA bandwidth. The transformer ratio is given by Equation (2).

$$n = \sqrt{\frac{2 \cdot R_{opt}}{50}} \qquad (2)$$

In Equation (2) R_{opt} is the optimum load of the main amplifier for delivering the maximum available power. Ideally speaking the impedance of combiner node could be decreased as low as possible, hence the optimum output power of Doherty is not restricted by usual bandwidth limitations. However, it needs to be noted that the ratio of the combining transformer should be kept in range of implementable value regarding the available technology for target application. Considering the practical restrictions, the modified topology in Figure 10 (b) shows a very wideband behaviour in block diagram level analysis and simulation.

In Table I the evaluated performances for different modifications are compared. P_{out} is the output power of DPA which is kept 19 dBm for all cases. BW of P_{out} is the BW of the output power level of the DPA that has lower than 0.5 dB variation over frequency and $\eta_{avg}(\%)$ is the BW average efficiency that is 2% lower than peak average efficiency.

V. CONCLUSION

A comprehensive review has been carried out on different modification of output combiner network for DPA for broadband transmitter application. Dependency of BW of PA to the output power level has been illustrated and the BW enhancement method is applied for different modification of Doherty output network in order to show the improvement. The BW of the maximum output power and average efficiency were evaluated as outcome to summarize. The presented systematic comparison will be helpful to PA designers to select the most promising topology of Doherty output combiner network for broadband wireless 5G transmitter.

TABLE I: Evaluated performances for different topologies

Topology	BW of $P_{out}(GHz)$	BW of $\eta_{avg}(GHz)$
Conventional DPA	3	1.5
Conventional DPA with BW enhancement method	3	2
Inverted DPA	4	3
Voltage-combined DPA	5.5	5
Voltage-combined with BW enhancement method DPA	7.5	7.5

ACKNOWLEDGEMENTS

The presented research work was done in "Front-End Integrated Circuits and Systems (FEIC)" research group at Silicon-Austria Labs (SAL) as a part of a cooperative project with Infineon Austria GmbH.

REFERENCES

[1] F. Wang, T. Li, S. Hu, and H. Wang, "A Super-Resolution Mixed-Signal Doherty Power Amplifier for Simultaneous Linearity and Efficiency Enhancement," *IEEE Journal of Solid-State Circuits*, vol. 54, no. 12, pp. 3421–3436, 2019.

[2] S. Hu, F. Wang, and H. Wang, "A 28-/37-/39-GHz Linear Doherty Power Amplifier in Silicon for 5G Applica-

tions," *IEEE Journal of Solid-State Circuits*, vol. 54, no. 6, pp. 1586–1599, 2019.

[3] G. Nikandish, R. B. Staszewski, and A. Zhu, "Breaking the Bandwidth Limit: A Review of Broadband Doherty Power Amplifier Design for 5G," *IEEE Microwave Magazine*, vol. 21, no. 4, pp. 57–75, 2020.

[4] P. M. Asbeck, "Will Doherty Continue to Rule for 5G?" in *2016 IEEE MTT-S International Microwave Symposium (IMS)*, 2016, pp. 1–4.

[5] J. Esch, "High-Efficiency Doherty Power Amplifiers: Historical Aspect and Modern Trends," *Proceedings of the IEEE*, vol. 100, no. 12, pp. 3187–3189, 2012.

[6] R. Pengelly, C. Fager, and M. Ozen, "Doherty's Legacy: A History of the Doherty Power Amplifier from 1936 to the Present Day," *IEEE Microwave Magazine*, vol. 17, no. 2, pp. 41–58, 2016.

[7] F. H. Raab, "Average efficiency of class-g power amplifiers," *IEEE Transactions on Consumer Electronics*, vol. CE-32, no. 2, pp. 145–150, 1986.

[8] S. Chen, G. Wang, Z. Cheng, and Q. Xue, "A Bandwidth Enhanced Doherty Power Amplifier with a Compact Output Combiner," *IEEE Microwave and Wireless Components Letters*, vol. 26, no. 6, pp. 434–436, 2016.

[9] E. Kaymaksut and P. Reynaert, "Transformer-Based Uneven Doherty Power Amplifier in 90 nm CMOS for WLAN Applications," *IEEE Journal of Solid-State Circuits*, vol. 47, no. 7, pp. 1659–1671, 2012.

[10] Z. Zong, X. Tang, K. Khalaf, D. Yan, G. Mangraviti, J. Nguyen, Y. Liu, and P. Wambacq, "A 28 GHz Voltage-Combined Doherty Power Amplifier with a Compact Transformer-based Output Combiner in 22nm FD-SOI," *IEEE Radio Frequency Integrated Circuits Symposium*, pp. 299–302, 2020.

[11] Y. Cho, K. Moon, B. Park, J. Kim, and B. Kim, "Voltage-Combined CMOS Doherty Power Amplifier Based on Transformer," *IEEE Transactions on Microwave Theory and Techniques*, vol. 64, no. 11, pp. 3612–3622, 2016.

[12] C. Zhao, B. Park, Y. Cho, and B. Kim, "Analysis and Design of CMOS Doherty Power Amplifier Using Voltage Combining Method," in *2013 IEEE International Wireless Symposium (IWS)*, 2013, pp. 1–4.

Design Procedure for a mm-Wave Transformer Based Matching Network with Spurious Rejection

Pankaj Venuturupalli*, Sina Mortezazadeh Mahani*, Franz Kuttner*,
Sondón Santiago Martin† and Johannes Sturm*
*Silicon Austria Labs GmbH, Villach, Austria
†Carinthia University of Applied Sciences, Villach, Austria
Email: pankaj.venuturupalli@silicon-austria.com

Abstract—This paper presents a design procedure for implementation of a matching network with spurious tones rejection using on-chip transformer for high speed DAC applications. Such circuits are complex to analyze and design, due to the higher order of the network. The presented design procedure involves matching the impedance at a desired center frequency based on the input impedance equation and with the aid of mathematical computational platform to avoid tedious mathematical derivations. A test case is considered with center frequency of 28GHz and detectable spurious tones at 30GHz and 26GHz to verify the proposed design procedure. The designed network matches a 50Ω load to a complex impedance of 12-j5.3 and simultaneously rejects the spurious tones by imposing notch filter at 30GHz and suppressing the other tone at 26GHz by 9.5dB.

Index Terms—Transformer based matching network, mm-Wave, non-linear spurs filtering, equation solving.

I. INTRODUCTION

The recent trend in the world of semiconductor industry has been to integrate sub-systems on to a single chip which has lead to the idea of integrating passive components such as transformers on to the chip [1]. Transformers are used as inter-stage networks for biasing the cascaded stages in addition to impedance transformation and differential to single ended conversion [2]. The characterization of a transformer based matching network is done based on the center frequency and impedance transformation ratio [3]. In this work, the design procedure for an impedance matching network to attain a certain impedance at the required center frequency is described, which also imbibes spurious tones rejection. In a high speed DAC the harmonic mixing between the clock and the signal results in spurious components at $mf_{CLK}\pm nf_o$ for all $m, n\epsilon Z$ [4]. The additional filtering can be used to filter out the nonlinear spurs that are generated near the desired signal.

This work aims to design and optimize the matching network in mm-Wave range for a 5G transmitter. The reported transformer based matching networks function at lower frequencies [6], [7] and are wide-band in mm-Wave range [8]–[11], which do not include impedance matching and spurious rejection simultaneously. Single LC tuned matching networks have straight forward design procedures implemented in [6] however, higher order networks are complicated to design [11]. This is the instant when mathematical computational platforms aid the designer to reach the specifications precisely without tedious mathematical derivations. The proposed methodology aimed at designing matching network with spurious rejection at 5G operating frequencies is observed to have validity in sub-GHz range too. The design procedure proposed is valid for other approaches such as in [6] and [11] with few alterations which are not discussed in this work. To the best of authors knowledge this is the first work describing a design and optimization procedure to design a transformer based impedance matching network simultaneously rejecting spurious tones using mathematical computational platforms.

The article is organized as follows: section II explains the design procedure steps, section III verifies the aforementioned design procedure considering a test case and followed by a conclusion in section IV

II. DESIGN PROCEDURE

Transformer based matching network along with spurious tones rejection is shown in Fig. 1 and its equivalent T-network is shown in Fig. 2. L_p and L_s are the primary and secondary inductors of the transformer respectively with a magnetic coupling coefficient k. C_1 shunts the secondary inductor which form a parallel resonant equivalent. The other parallel resonant pair L_n, C_n placed in series performs the spurious tones rejection at ω_n given by eq. 1.

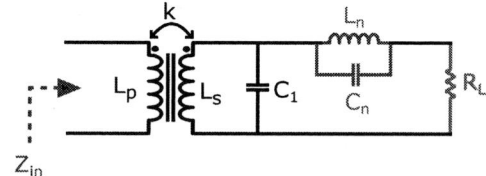

Fig. 1: Transformer matching network with a notch filter

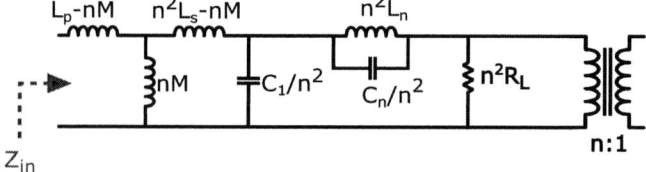

Fig. 2: Circuit model based on transformer T-model used for mathematical derivation

$$\omega_n^2 = \frac{1}{L_n C_n} \qquad (1)$$

978-1-7281-8494-4/20 $31.00 © 2020 IEEE

The aim of the circuit is to transform the load resistance R_L to a desired impedance at the frequency of interest. The complex conjugate matching with the real part being identical to the preceding stage leads to maximum power transfer with an efficiency of 50%. Efficiency can be maximized by considering the real part of the input impedance greater than that of preceding stage [12] by a factor α as reported in [6]. The frequency dependent input impedance of the equivalent transformer T-model shown in Fig. 2 is given by eq. 2 and the real part and the imaginary part are given by eq. 3 and eq. 4 respectively.

$$Z_{in} = \frac{N_r + jN_i}{D_r + jD_i} \tag{2}$$

$$Re(Z_{in}) = \frac{N_r D_r + N_i D_i}{D_r{}^2 + D_i{}^2} \tag{3}$$

$$Im(Z_{in}) = \frac{N_i D_r - N_r D_i}{D_r{}^2 + D_i{}^2} \tag{4}$$

where

$$N_r = \omega^2((M^2 - L_p L_s)(1 - \omega^2 L_n C_n - \omega^2 L_n C_1) - L_p L_n)$$

$$N_i = \omega R_L(1 - \omega^2 L_n C_n)(L_p + \omega^2 C_1(M^2 - L_p L_s))$$

$$D_r = R_L(1 - \omega^2 L_n C_n)(1 - \omega^2 L_s C_1)$$

$$D_i = \omega L_n + \omega L_s(1 - \omega^2 L_p C_n - \omega^2 L_n C_1)$$

The design procedure is given as a flowchart shown in Fig. 3. The input frequency of the notch is to be defined according to the occurrences of the nonlinear spurs. The selection of the capacitors C_1 and C_n affects the performance such as bandwidth and the magnetic coupling to attain a certain impedance value as shown in Fig. 4 and Fig. 5. The design procedure comprises of 3 major steps: Firstly, fixing the value of L_n based on the selected series resonance value. Secondly, calculating the value of L_s to tune the matching network to the desired center frequency f_c and finally, calculating L_p, k to set the input impedance.

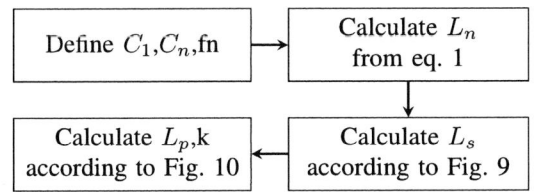

Fig. 3: Design procedure of the matching network with spurious tones rejection

The conventional method of finding the center frequency is equating the imaginary part of the input impedance to zero, but for complex conjugate matching the imaginary part cannot be zero rather it should be a specific value. The

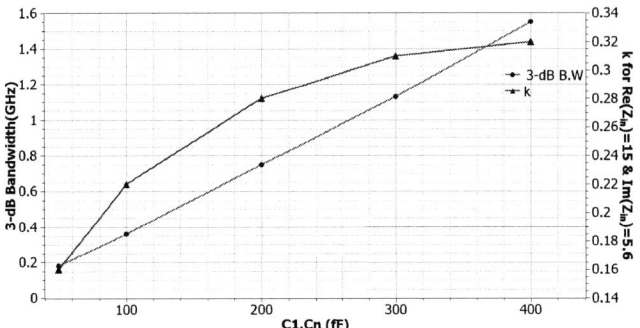

Fig. 4: Variation of 3-dB bandwidth and k with equal C_1, C_n

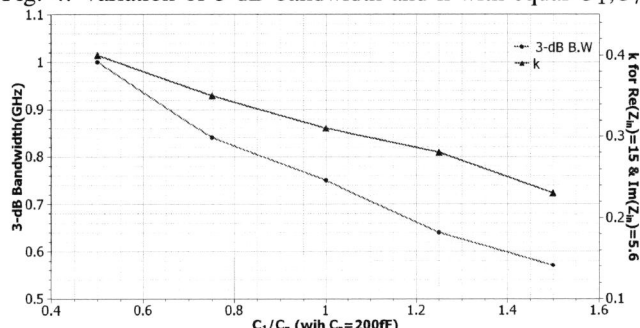

Fig. 5: Variation of 3-dB bandwidth and k with unequal C'_1, C'_n

procedure involves finding the imaginary terms that cancel out at the frequency of interest and then deriving the mathematical equation for center frequency as stated in [6]. With the higher order of the network it becomes a tedious task to calculate the center frequency mathematically. The procedure for selecting the value of L_s which in turn decides f_c is given as a flowchart shown in Fig. 9. L_s is to be independent of selected L_p and k, for this to be satisfied L_p should be in the order of few pH as shown Fig. 6, and k should be low as shown in Fig. 7. The range specified is where L_s is independent of L_p and k.

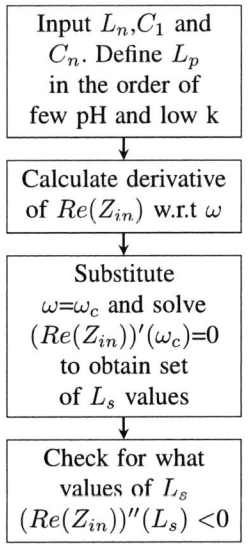

Fig. 9: Calculating L_s to fix the center frequency

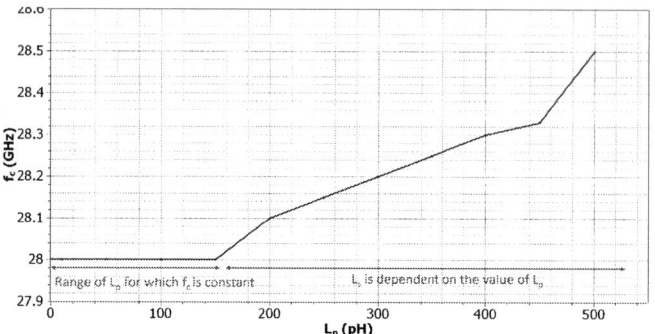

Fig. 6: Range of L_p for a constant f_c

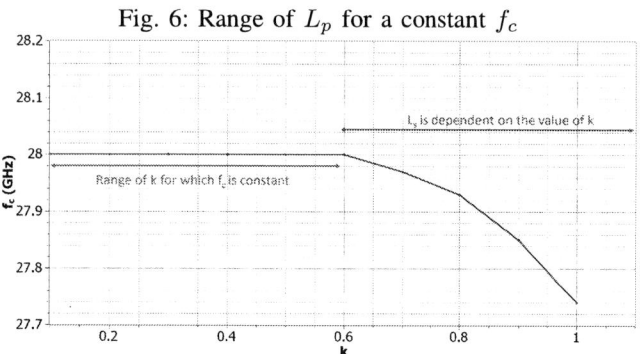

Fig. 7: Range of k for a constant f_c

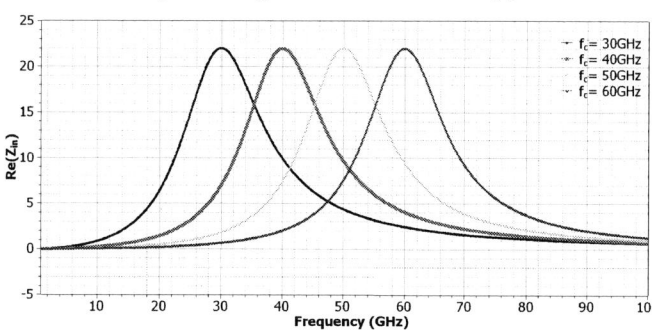

Fig. 8: Real part of Z_{in} peak at center frequency

The real part of input impedance peaks at the center frequency as shown in Fig. 8 for center frequencies of 30 to 60GHz. This tones up the idea of using maxima finder of a function using first and second order derivatives as depicted in Fig. 9. The procedure of finding maxima of a function involves solving the first order derivative with respect to ω and equating to zero. With all the remaining values specified(L_p, k, C_1, L_n, C_n), the unknowns in the solved derivative equation are ω and L_s. Now, substituting the desired frequency in the equation leads to a set of values for L_s by solving $Re(Z_{in})'(\omega_c)=0$. To find if a specific value is where the function attains maxima value, the second order derivative must be calculated and at the specific value the second order derivative must be negative. The L_s which solves this criteria is the desired value.

With the network tuned to the frequency of interest, the next step would be to set the values of L_p and k such that the desired input impedance can be achieved. The procedure for

finding out the appropriate L_p and k is shown in Fig. 10 which is iterative process. L_p and k are initiated by equating to 1pH and 0.01 respectively. A conditional statement is evaluated to check if the obtained values of $Re(Z_{in})$ and $Im(Z_{in})$ by substituting L_p and k is equal to the desired values $Re(Z_{in})^*$ and $Im(Z_{in})^*$. The values of L_p and k for which the desired values are attained are tabulated within in a error range of $\pm 2\%$.

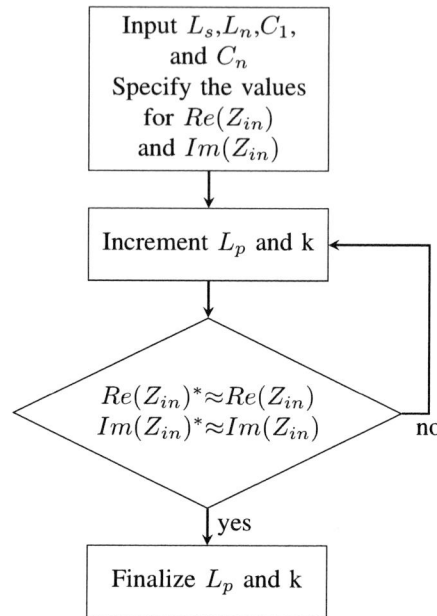

Fig. 10: Calculating L_p and k

III. IMPLEMENTATION

The proposed design procedure is verified by considering a test case according to selected mm-Wave bands for 5G mobile communication, operating at a center frequency of 28GHz [5]. At this frequency, the design procedure for a transformer based matching network with spurious rejection is verified to transform a 50Ω load to a complex impedance of $12-j5.3$ and spurious tone at 30GHz. For verification purpose of the design procedure the passive components are assumed to have a high quality factor. For the reasons explained in Section II the real part is matched to 15Ω. From the analysis shown in Fig. 5, capacitance of C_1 and C_n are taken to be 100fF and 200fF to have a higher magnetic coupling. Upon following the procedure mentioned earlier, the values of the matching network are tabulated in the table I. The S_{21} and the impedance variation with frequency is shown in Fig. 11, the rejected spurious mode is 30GHz, the other mode at 26GHz has an attenuation of 9.5dB. The attenuation at the center frequency 28GHz is 1.46dB. The real part has a value of 15Ω which also turns out to be the maxima value of the function $Re(Z_{in})$ which occurs at the center frequency.

TABLE I: Generated values for $f_c=28GHz$

f_c	f_n	k	L_p	L_s	L_n	C_1	C_n
28GHz	30GHz	0.4	57.3pH	455.23pH	140.72pH	100fF	200fF

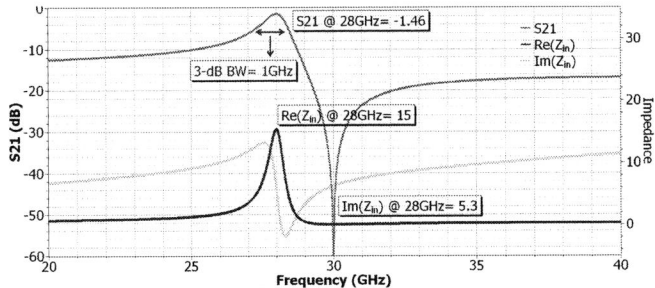

Fig. 11: S21 and impedance plot for f_c=28GHz

Comparison between a notch filter and a notch-less filter is shown in Fig. 12, it is clear that without the notch the circuit has much wider bandwidth and the nonlinear spurious have lower rejection as tabulated in table II.

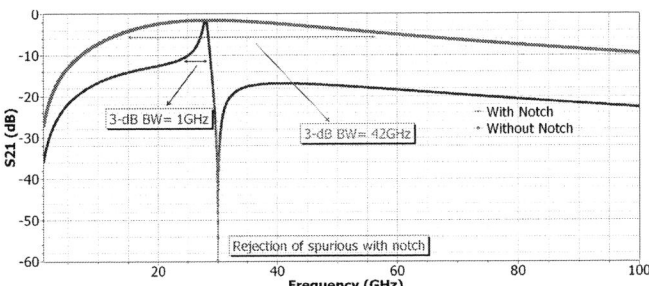

Fig. 12: S21 plot for a notch based and notch-less MN for f_c=28GHz

TABLE II: Comparison between notch and notch-less response for f_c=28GHz

Parameter	Notch	Notch-less
@28GHz	-1.46dB	-1.5dB
@30GHz	Notch	-1.53dB
@26GHz	-9.5dB	-1.6dB
3-dB bandwidth	1GHz	42GHz

IV. CONCLUSION

Design procedure for a transformer based matching network with simultaneous spurious tone rejection is presented in this paper. The use of mathematical computational platform aids the designer by avoiding tedious mathematical derivations and also in optimizing in terms of tuning and impedance matching. The design procedure is verified for a test case with center frequencies of 28GHz and spurious tones at 30GHz matched to a complex impedance $12 - j5.3$. Comparison with a notch-less design has been made which shows a difference of 8dB higher suppression at the side where the notch is not present and much higher suppression at the notch frequency.

ACKNOWLEDGMENT

This work has been done in the RF Frontend Lab, jointly supported by FH Kärnten (CUAS) and by Silicon Austria Labs (SAL), owned by the Republic of Austria, the Styrian Business Promotion Agency (SFG), the federal state of Carinthia, the Upper Austrian Research (UAR), and the Austrian Association for the Electric and Electronics Industry (FEEI).

REFERENCES

[1] J. R. Long, "Monolithic transformers for silicon RF IC design," in IEEE Journal of Solid-State Circuits, vol. 35, no. 9, pp. 1368-1382, Sept. 2000.

[2] D. Chowdhury, P. Reynaert and A. M. Niknejad, "Design Considerations for 60 GHz Transformer-Coupled CMOS Power Amplifiers," in IEEE Journal of Solid-State Circuits, vol. 44, no. 10, pp. 2733-2744, Oct. 2009.

[3] A. R. Suvarna, V. Bhagavatula and J. C. Rudell, "Transformer-Based Tunable Matching Network Design Techniques in 40-nm CMOS," in IEEE Transactions on Circuits and Systems II: Express Briefs, vol. 63, no. 7, pp. 658-662, July 2016.

[4] S. Balasubramanian et al., "Systematic Analysis of Interleaved Digital-to-Analog Converters," in IEEE Transactions on Circuits and Systems II: Express Briefs, vol. 58, no. 12, pp. 882-886, Dec. 2011.

[5] H. Hashemi, "Millimeter-wave power amplifiers and transmitters," 2017 IEEE Custom Integrated Circuits Conference (CICC), Austin, TX, 2017, pp. 1-8.

[6] A. Eldeeb, G. Batistell, W. Bösch and J. Sturm, "SCPA Transformer Based Matching Network Design Flow and SiP Implementation," 2018 Austrochip Workshop on Microelectronics (Austrochip), Graz, 2018, pp. 22-26, doi: 10.1109/Austrochip.2018.8520704.

[7] G. Batistell, T. Holzmann, S. Leuschner, A. Wolter, A. Passamani and J. Sturm, "SiP solutions for wireless transceiver impedance matching networks," 2017 12th European Microwave Integrated Circuits Conference (EuMIC), Nuremberg, 2017, pp. 329-332, doi: 10.23919/Eu-MIC.2017.8230725.

[8] Gomha Siddig, Langat Kibet. (2018). On-chip 60 GHz transformers for next generation of wireless communication in a 0.13μ BiCMOS technology. 1-4. 10.1109/ICTAS.2018.8368760.

[9] M. S. Hossain, M. Fujishima, T. Yoshida, S. Amakawa and M. M. Rashid, "Design of CMOS On-Chip Transformer Coupled Matching Network for Millimeter-Wave Amplifiers with Optimal Chip Area," 2019 1st International Conference on Advances in Science, Engineering and Robotics Technology (ICASERT), Dhaka, Bangladesh, 2019, pp. 1-6, doi: 10.1109/ICASERT.2019.8934574.

[10] G. Jin, Y. Kong, H. Min, H. Xu and Y. Na, "Transformer based broadband matching," 2019 Cross Strait Quad-Regional Radio Science and Wireless Technology Conference (CSQRWC), Taiyuan, China, 2019, pp. 1-3, doi: 10.1109/CSQRWC.2019.8799250.

[11] A. Mazzanti and A. Bevilacqua, "Second-Order Equivalent Circuits for the Design of Doubly-Tuned Transformer Matching Networks," in IEEE Transactions on Circuits and Systems I: Regular Papers, vol. 65, no. 12, pp. 4157-4168, Dec. 2018, doi: 10.1109/TCSI.2018.2846029.

[12] M. N. Abdallah, T. K. Sarkar and M. Salazar-Palma, "Maximum power transfer versus efficiency," 2016 IEEE International Symposium on Antennas and Propagation (APSURSI), Fajardo, 2016, pp. 183-184, doi: 10.1109/APS.2016.7695800.

Edge-Combining for QPSK and FSK Modulation Using a Single Delay-Line

Markus Stadelmayer, *Student Member, IEEE*, Tim Schumacher, *Student Member, IEEE*, Thomas Faseth, *Member, IEEE*, Oliver Lang, *Member, IEEE*, Harald Pretl, *Senior Member, IEEE*

Johannes Kepler University Linz, Austria

Email: see http://www.iic.jku.at/analog/team

Abstract—This work introduces a method to use edge-combining to obtain Quadrature Phase Shift Keying (QPSK) as well as Frequency Shift Keying (FSK) signals in a power efficient way to be used in low-power integrated transmitters for Internet of Things applications. Therefore, 16 time-delayed (phase-shifted) signals are produced by one 8-stage differential delay line. The signals are applied in combination with edge-combining for both, 4-times frequency multiplication and quadrature signal generation. Moreover, a phase selector is assigned to perform QPSK as well as FSK by adapting the selector stepping values according to the desired frequency deviation. In order to get better spectral purity, additional phase values (as needed for 12-PSK) for frequency modulation are generated by additional phase shifts. For this work, the concept is proven by simulation and targets a phase locked ring oscillator with a constant in-band phase noise of −80 dBc/Hz in the loop bandwidth of 1 MHz and a out of band phase noise with a slope of −20 dB/dec from a carrier frequency of 864 MHz. The proposed method intends to reduce the power demand of integrated transmitters by keeping the frequency and amount of power-hungry components low. Moreover, compared to other work in the area of edge-combining transmitters, a method for complex modulation by one delay line is introduced.

Index Terms—Delay-line, edge-combining, FSK, low-power, QPSK, transmitter.

I. INTRODUCTION

IN recent years the Internet of Things (IoT) obtains more and more importance [1]. It consists of many autonomous devices that are connected with each other—most of them using a wireless data link. As many of the devices are battery-powered, they need to have a low power demand to extend battery life and charging cycles. Usually, the most power-hungry part of an IoT-device is the transmitter as it contains a power-consuming oscillator and power amplifier.

Classical transmitter designs use IQ-mixing architectures to provide efficient coherent modulation schemes but simultaneously have a high power demand due to their complexity as well as component count. In order to reduce the power demand, much effort [2–6] has been spent in recent years as [7] summarizes. Some architectures minimize the number of components in the transmit path. E.g. one of the quadrature paths can be omitted with the drawback that only non-complex modulation schemes like Binary Phase Shift Keying (BPSK) or On Off Keying (OOK) can be performed. Moreover, the mixer of the remaining path can be omitted by running an oscillator directly at radio frequency [3].

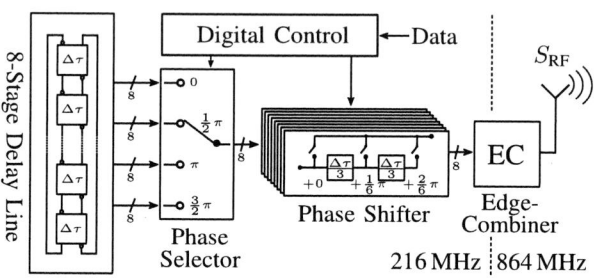

Fig. 1. Block diagram of the proposed selection modulator.

Another approach to reduce power demand is to lower the operating frequency of the transmitter components. Therefore, the concept of edge-combining (Sec. II) is applied to decrease the oscillator and phase locked loop frequency. This is achieved by performing frequency multiplication to generate the radio frequency signal [2, 6, 8–10]. Moreover, [11] has shown that it can nearly half the power demand in comparison to run the oscillator directly at radio frequency.

In this work, a method is proposed to utilize edge-combining to reduce power demand but still provide complex modulation capabilities like Quadrature Phase Shift Keying (QPSK) and Frequency Shift Keying (FSK). As pictured in Fig. 1, the modulator uses a single 8-stage delay line—realized as a differential ring oscillator—running at merely 216 MHz to provide 16 time-delayed (phase-shifted) signals. Groups of 8 signals are switched to the edge-combiner by the phase selector for both, QPSK modulation and 4-times frequency multiplication to hit the 864 MHz ISM-band. Moreover, additional phase shifts are obtained by extra delay lines (phase shifter) to generate additional phases as also used for 12-PSK. By adapting the phase selector and phase shifter stepping values according to the desired frequency deviation the digital frequency modulation is performed. Besides low-power operation and providing complex modulation capabilities, the method enables to build up a fully integrated low-power transmitter in a small area, as no inductors are needed. Hence, it is suitable for IoT-applications.

978-1-7281-8494-4/20 $31.00 © 2020 IEEE

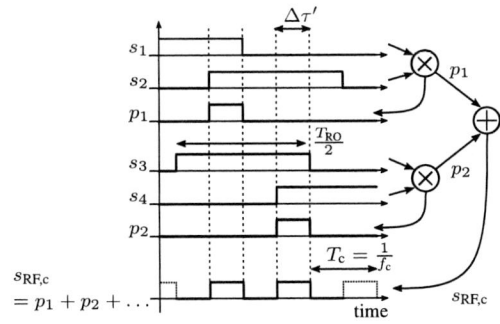

Fig. 2. Theoretical signal generation using edge-combining.

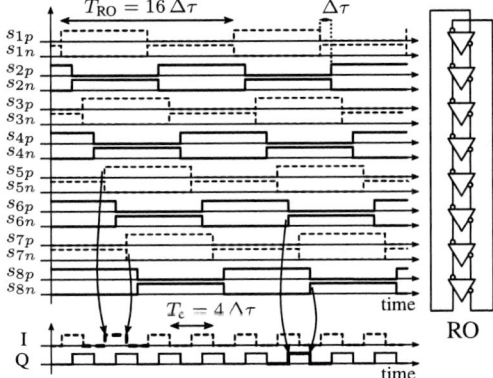

Fig. 3. Edge-combining using an 8-stage differential ring oscillator (RO) for quadrature signal generation.

II. CARRIER GENERATION

The radio frequency carrier $s_{\text{RF,c}}$ is generated by using the edge-combining principle. The idea is to create a high frequency $f_c = 1/T_c$ signal from a signal with a lower frequency $f_{\text{RO}} = 1/T_{\text{RO}}$ to save power by reducing the operating frequency. Therefore, several time-delayed (phase-shifted) signals are generated by a delay line and connected with logical operations as described in the following.

By multiplying two delayed signals s_1 and s_2 (s_3 and s_4), a pulse p_1 (p_2) can be generated at the time $\Delta\tau'$ when those signals overlap as shown in Fig. 2. If several of these pulses (remaining pulses marked as dotted line) are added up $p_1 + p_2 + \cdots$, a radio frequency signal $s_{\text{RF,c}}$ with a frequency of $f_c = 1/(2\Delta\tau')$ is created. The time-delayed signals are generated from a delay line running at a low frequency f_{RO}.

The proposed design uses an 8-stage ($K = 8$) pseudo differential ring oscillator with inverted feed-back as delay line. It is intended to be regulated by a phase locked loop as described in [12]. The oscillator provides $2 \cdot K = 16$ time-delayed (phase-shifted) signals ($s_{k\{p|n\}}$ with $k \in \{1, 2, \ldots, 8\}$) pictured in Fig. 3. Each stage has a delay of $\Delta\tau$, causing a ring oscillator frequency of $f_{\text{RO}} = 1/T_{\text{RO}} = 1/(2K\Delta\tau)$. By applying $M \in \{2, 4, 8\}$ of the K stages for edge combining the delay line frequency f_{RO} is multiplied with the frequency multiplication factor of $m = M$. Hence, a carrier signal $s_{\text{RF,c}}$ with a frequency of $f_c = m \cdot f_{\text{RO}}$ is generated.

TABLE I
CONSTELLATION TABLE EDGE-COMBINING

0	$\pi/2$	π	$3\pi/2$
$s_{1p} \cdot s_{3n}$	$s_{2n} \cdot s_{4p}$	$s_{3p} \cdot s_{5n}$	$s_{4n} \cdot s_{6p}$
$+s_{5p} \cdot s_{7n}$	$+s_{6n} \cdot s_{8p}$	$+s_{7p} \cdot s_{1p}$	$+s_{8n} \cdot s_{2n}$
$+s_{1n} \cdot s_{3p}$	$+s_{2p} \cdot s_{4n}$	$+s_{3n} \cdot s_{5p}$	$+s_{4p} \cdot s_{6n}$
$+s_{5n} \cdot s_{7p}$	$+s_{6p} \cdot s_{8n}$	$+s_{7n} \cdot s_{1n}$	$+s_{8p} \cdot s_{2n}$

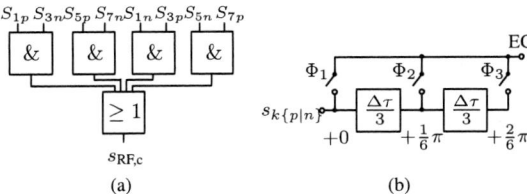

Fig. 4. Block diagram of (a) the edge-combiner and (b) the phase shifter.

The maximum frequency multiplication factor $m_{\max} = 8$ is obtained by using the signals from all of the K stages ($m_{\max} = M_{\max} = K$) to generate the radio frequency signal $S_{RF,c}$. Thus, a radio frequency of $f_{c,\max} = K f_{\text{RO}}$ is derived. In this case, the signals of two subsequent stages are multiplied pairwise and the resulting pulses are added up. Consequently, the maximum achievable frequency is just dependent on the delay $\Delta\tau$ of one delay cell as it represents one-half-wave of the radio frequency signal and is calculated with $f_{\text{RF,max}} = 1/(2\Delta\tau)$. Hence, the maximum output frequency is limited by the oscillator speed/delay itself and does not rise with the number of stages.

The proposed architecture utilizes a frequency multiplication factor of $m = 4$ as thereby quadrature signals, needed for modulation (see Sec. III), can be derived. In this case, the signals of every second delay stage are multiplied with each other in pairs to produce either the in-phase I ($\varphi = 0$) or quadrature Q ($\varphi = \pi/2$) component. A carrier frequency of $f_c = 1/T_c = 1/(4\Delta\tau)$ is produced by adding them up. As already mentioned, Fig. 3 illustrates the node voltages of the ring oscillator marked as $s_{k\{p|n\}}$. The eight dashed signals are used for generating the in-phase I component by, for example, multiplying s_{5p} with s_{7n}. Moreover, the eight solid signals represent the quadrature Q component (e.g. s_{6n} multiplied with s_{8p}). In Tab. I the signals that have to be multiplied pairwise and added up in order to generate the in-phase and quadrature components are summarized. Moreover, the combination used to create the inverse signals $\varphi = \pi$ and $\varphi = 3\pi/2$ are given.

In hardware, the edge-combining operation is performed by the structure pictured in the block diagram of Fig. 4(a) with exemplary signals for the in-phase component. The multiplication that is executed in Fig. 2 and Tab. I is achieved by the AND-gates and the summation by the OR-gate. Several other implementations are demonstrated in [11].

III. MODULATION

Previous section (Sec. II) describes the generation of quadrature signals by using edge-combining. Based on that,

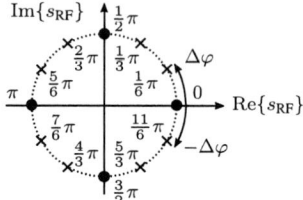

Fig. 5. Constellation diagram of QPSK and 12-PSK points.

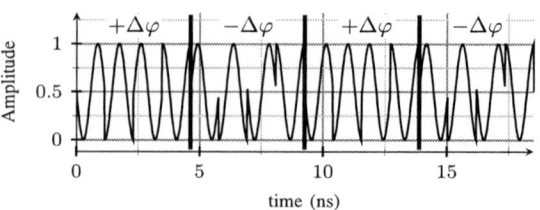

Fig. 6. Frequency modulated signal for $\Delta\varphi = \pm\pi/2$ by switching every period.

this section explains how to modulate data on the resulting radio frequency carrier. As briefly described in Sec. I, for that purpose the architecture shown in Fig. 1 is used for phase and frequency modulation.

A. Phase Modulation

Digital phase modulation is done via a selection modulator. It is realized as a multiplexer, that is able to switch signals having different phases representing digital data words to the antenna. As described in Sec. II, groups of eight signals can be formed within the edge-combiner to a radio frequency signal having one of the QPSK phases ($\varphi \in \{0, \pi/2, \pi, 3\pi/2\}$). The block diagram in Fig. 1 shows that the phase selector block is able to select one of the groups and switch it to the phase shifter and edge-combiner. If the phase shifter is bypassed, the QPSK points marked as dots in Fig. 5 can be formed by the edge-combiner. Moreover, the QPSK points are extended to the points of 12-PSK (used for frequency modulation and marked as crosses in Fig. 5) by applying an additional phase shift within the phase shifter pictured in Fig. 4(b). This phase shifter is realized by two $\Delta\tau/3$ time delay elements connected in cascade. Each delay results in a phase shift of $\Delta\varphi = \pi/6$ of the transmit signal ($T_c = 1/f_c = 4\,\Delta\tau$, Sec. II).

In hardware the delay can be realized by inverter gates that are matched to the inverters of the ring oscillator. For proper matching it is important that the scaling between a ring oscillator stage and a phase-shifter stage is not large. In this architecture a scaling factor of three for a delay of $\Delta\tau/3$ one phase shifter stage is chosen. As mentioned before, if the phase shifter is bypassed by closing switch Φ_1 then the points $\varphi \in \{0, \pi/2, \pi, 3\pi/2\}$ will be produced. Moreover, closing Φ_2 results in an additional phase shift of $\pi/6$ and the constellation points $\varphi \in \{\pi/6, 4\pi/6, 7\pi/6, 10\pi/6\}$ are achieved. Likewise, by closing Φ_3 an alternate phase shift of $2\pi/6$ and the constellation points $\varphi \in \{2\pi/6, 5\pi/6, 8\pi/6, 11\pi/6\}$ are created.

The switches, and correspondingly the modulation (phase selector and phase shifter), are controlled by a digital control block (Fig. 1). The symbol rate f_S equals the switching frequency f_{sw} of the phase selector and the phase shifter with a corresponding data rate of $DR_{QPSK} = 2 \cdot f_S$.

B. Frequency Modulation

A binary digital frequency modulation with a frequency deviation of $\pm f_{mod}$ is emulated by repeatedly changing the signal phase $\varphi(t)$ of the transmit signal $s_{RF}(t)$ in steps. A positive frequency deviation $+f_{mod}$ from the carrier frequency f_c is obtained by increasing the phase with a step size of $+\Delta\varphi$ and a switching frequency of $f_{sw} = 1/T_{sw}$. Decreasing the phase ($-\Delta\varphi$) results in a negative frequency deviation $-f_{mod}$ as given in (1). Fig. 6 exemplary shows frequency modulated signal by changing the phase in every signal period by $\Delta\varphi = \pm\pi/2$.

$$\pm f_{mod} = \frac{1}{2\pi} \frac{\pm\Delta\varphi}{T_{sw}} = \frac{\pm\Delta\varphi}{2\pi} f_{sw} \tag{1}$$

The phase steps $\Delta\varphi$ are received from the selection modulator using the phases of the digital phase modulation explained in Sec. III-A ($\Delta\varphi = \pi/2$, QPSK and $\Delta\varphi = \pi/6$). In order to mathematically describe the phase stepped transmit signal $s_{RF}(t)$, initially, a signal with a general time varying phase $\varphi(t)$ at a carrier frequency of f_c is considered in (2):

$$s_{RF}(t) = \cos(2\pi f_c t + \varphi(t)). \tag{2}$$

Assuming that $s_{RF}(t)$ approximates a bandpass signal, it can be transferred to the baseband $s_{BB}(t)$ to just observe its phase behavior and neglect the carrier:

$$s_{BB}(t) = e^{j\varphi(t)}. \tag{3}$$

The phase steps are modeled in (4) by multiplying a baseband signal $\exp[j(\pm\Delta\varphi\,t/T_{sw})]$ without steps by a sum of dirac pulses represented as Dirac-comb with a frequency of $f_{sw} = 1/T_{sw}$. Hence, the values at the stepping points $t = \kappa T_{sw}$ ($\varphi = \kappa\Delta\varphi$) are derived. The rectangular shape of the phase steps is then described by convoluting with a rectangular signal (duration T_{sw}):

$$s_{BB}(t) = \text{rect}\left(\frac{t}{T_{sw}}\right) * \left(e^{j(\pm\Delta\varphi\,t/T_{sw})} \sum_{\kappa=-\infty}^{\infty} \delta(t - \kappa\,T_{sw})\right). \tag{4}$$

The corresponding signal in frequency domain $S_{BB}(f)$, where T_{sw} is replaced by $1/f_{sw}$, is given in (5). The multiplication with the Dirac-comb transforms into a convolution and the convolution with rectangular signal corresponds to a multiplication with $\text{si}(x) = \sin(x)/x$, scaled with $1/f_{sw}$ due to the rectangular shaped phase steps:

$$
\begin{aligned}
&S_{BB}(f) \\
&= \frac{1}{f_{sw}}\text{si}\left(\pi\frac{f}{f_{sw}}\right)\left[\delta\left(f - \frac{\pm\Delta\varphi}{2\pi}f_{sw}\right) * \sum_{\kappa=-\infty}^{\infty}\delta(f - \kappa\,f_{sw})\right] \\
&= \frac{1}{f_{sw}}\text{si}\left(\pi\frac{f}{f_{sw}}\right)\left[\sum_{\kappa=-\infty}^{\infty}\delta\left(f - \kappa\,f_{sw} - \frac{\pm\Delta\varphi}{2\pi}f_{sw}\right)\right].
\end{aligned}
\tag{5}
$$

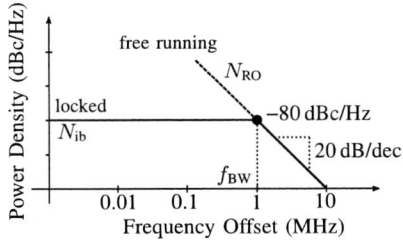

Fig. 7. Modelled phase noise of the phase locked delay line (ring oscillator)

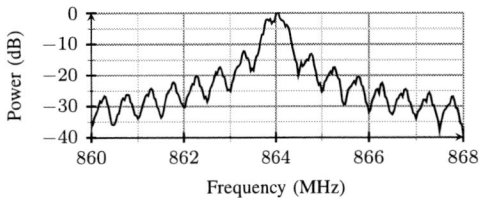

Fig. 8. Spectrum of the QPSK modulated signal.

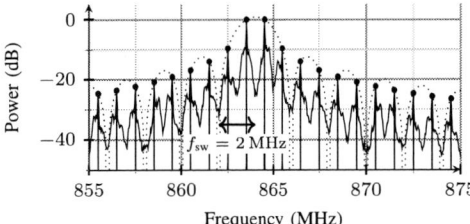

Fig. 9. Simulated (–) and calculated (●) FSK-spectrum ($f_{\text{mod}} = 500\,\text{kHz}$) generated with phase steps of $\Delta\varphi = \pi/2$.

By up-converting the baseband signal $S_{\text{BB}}(f)$ calculated in (5) to the carrier frequency f_c and inserting (1), the two-sided frequency modulated transmit signal $S_{\text{RF}}(f)$ is derived in (6):

$$
\begin{aligned}
S_{\text{RF}}(f) \\
= \frac{1}{2f_{\text{sw}}} &\left[\operatorname{si}\left(\pi \frac{f - f_c}{f_{\text{sw}}} \right) \left(\sum_{\kappa=-\infty}^{\infty} \delta\left(f - f_c - \kappa\, f_{\text{sw}} \pm f_{\text{mod}} \right) \right) \right. \\
&\left. + \operatorname{si}\left(\pi \frac{-f - f_c}{f_{\text{sw}}} \right) \left(\sum_{\kappa=-\infty}^{\infty} \delta\left(-f - f_c - \kappa\, f_{\text{sw}} \pm f_{\text{mod}} \right) \right) \right].
\end{aligned}
$$

$$(6)$$

The calculated spectrum (see Sec. IV, Fig. 9 and Fig. 10) consists of Dirac pulses that are weighted by the si-function centered around the carrier frequency f_c. The spectrum has the the desired peaks of binary frequency modulation at $f_c \pm f_{\text{mod}}$ for $\kappa = 0$ but also spectral repetitions at $\kappa \cdot f_{\text{sw}}$ caused by the phase stepping. For implementation a additional analog filter need to be applied in order to suppress the unwanted signal repetitions.

IV. RESULTS AND COMPARISON TO THE STATE-OF-THE-ART

The proposed method for applying complex-valued modulation formats to edge-combining transmitters is analyzed and verified using simulation (matlab, RF-toolbox). Therefore, the central element—the selection modulator—applying QPSK and FSK is modeled neglecting the delay mismatch in the delay line. A phase locked ring oscillator [12] running at $f_{\text{RO}} = 216\,\text{MHz}$ and having a loop bandwidth of $f_{\text{BW}} = 1\,\text{MHz}$ is assumed to be used. It causes a constant in-band (frequency offset: $f < f_{\text{BW}}$) phase noise $N_{\text{ib}} = -80\,\text{dBc/Hz}$ at the carrier frequency of $864\,\text{MHz}$ as pictured in Fig. 7. The out-of-band ($f > f_{\text{BW}}$) phase noise is dominated by the phase noise of the oscillator N_{RO}. It is modeled with a slope of $-20\,\text{dB/dec}$ and a power density of $-80\,\text{dBc/Hz}$ at the loop bandwidth corner frequency $f_{\text{BW}} = 1\,\text{MHz}$ (see Fig. 7). Integrating the phase noise from $1\,\text{kHz}$ to $10\,\text{MHz}$ as demonstrated in [13] leads to a root mean square phase error of $\varphi_{\text{RMS}} = 0.062\pi$. Therewith $8 \cdot \varphi_{\text{RMS}} = 0.496\pi < \pi/2$ indicates that $99.99\,\%$ of the symbols are transmitted in the right quadrant of the constellation diagram (Fig. 5).

As already mentioned the carrier signal is located at a frequency of $864\,\text{MHz}$ by having a frequency multiplication factor of $m = 4$. The resulting normalized spectrum of QPSK modulation and having a symbol rate of $f_S = 500\,\text{kHz}$ is plotted in Fig. 8. Moreover, frequency modulation is simulated by a transient baseband simulation with a frequency deviation of $f_{\text{mod}} = 500\,\text{kHz}$. This results in a switching frequency of $f_{\text{sw}} = 2\,\text{MHz}$ for $\Delta\varphi = \pi/2$ and $f_{\text{sw}} = 6\,\text{MHz}$ for $\Delta\varphi = \pi/6$ according to (1). The spectrum of the FSK signal $S_{\text{RF}}(f)$ for stepping with $\Delta\varphi = \pi/2$ and $\Delta\varphi = \pi/6$ are plotted in Fig. 9 and Fig. 10, respectively. Additionally, the calculated spectrum derived from (6)—consisting of Dirac pulses that are weighted by the si-function indicated as dotted line—is plotted as dots. The calculation fits the simulated points but the simulation additionally considers phase noise.

As derived in Sec. III-B, spectral repetitions can be observed at distances of $\kappa \cdot f_{\text{sw}}$. Changing the switching frequency leads to a scaling of the si-function and therewith the Dirac pulses in direction of the frequency axis. Moreover, keeping the switching frequency constant and changing $\Delta\varphi$ and f_{mod} moves the frequency points on the si-function marked as the dotted line.

The spectral repetitions can cause disturbances in other frequency bands. Therefore, they need to be suppressed as good as possible by a filter. In order to lower the filter requirements it is desirable to have a high switching frequency f_{sw} in combination with small phase steps $\Delta\varphi$ to move the unwanted peaks of the spectral repetitions farther from the carrier.

The proposed method is compared to other works targeting integrated low-power transmitter architectures with a focus on edge-combining and phase selection to be used in short-range IoT applications.

The publications [2, 6, 11] use edge-combining for frequency multiplication, but FSK modulation is either generated by manipulating the phase locked loop or the reference frequency. Hence, they use edge combining just for frequency multiplication but do not take advantage to additionally perform complex QPSK modulation.

978-1-7281-8494-4/20 $31.00 © 2020 IEEE

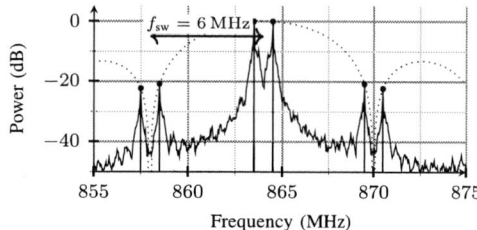

Fig. 10. Simulated (–) and calculated (•) FSK-spectrum ($f_{\text{mod}} = 500\,\text{kHz}$) generated with phase steps of $\Delta\varphi = \pi/6$.

Other reported works [5, 10] utilize different phases obtained from a delay line or divider in order to perform phase modulation using a multiplexer. These designs suffer from a high power demand by running the oscillator directly at radio frequency or even higher and obtain phase modulation by a divider. However, [8] uses edge-combining not only for frequency multiplication but also for BPSK-modulation. This is done by inverting the output signal. Moreover, [9] applies two different oscillators for generating quadrature signals to be used together with a phase selector. Therewith, a complex modulation is obtained. However, using two oscillators can lead to problems by synchronizing them with a phase deviation of $\pi/2$.

Compared to the others, in this work, the modulated signals are obtained from just a single delay line. This maintains a fixed relation of $\pi/2$ between the quadrature signals as it is only dependent on the delay of the stages which can be matched properly. Moreover, this work goes one step further and uses the quadrature signals not only for QPSK-modulation but also for emulating FSK modulation by adapting the selector stepping values according to the frequency deviation. Additional phases for 12-PSK are generated by further delay cells. With doing so, the spectral behavior of the FSK is improved. The proposed concept uses additional capabilities of edge-combining to have a wide modulation spectrum but simultaneously having a low power demand.

V. CONCLUSION

A method to generated a QPSK and FSK modulated radio frequency signal in a power efficient way using the concept of edge-combining from one delay line is introduced and the mathematical derivation is explained. Power is saved by minimizing the number of components and their operating frequency. Compared to other state-of-the-art ultra low power transmitters, the proposed approach additionally supports complex modulation formats. Moreover, as no inductors are used for the modulator, the method can be integrated into a small area. Hence, the concept not only saves power but also costs.

In further steps, the modulator is planned to be practically verified in a CMOS-process. As the transmitter blocks consists mainly of digital components, it will be scalable to different technology nodes. Therefore, it can be used flexibly in several applications where low power demand and low costs are needed. Examples for that are wireless sensor nodes and Internet of Things applications.

REFERENCES

[1] Ericsson, *Ericsson Mobility Report, November 2019*, 2019. [Online]. Available: https://www.ericsson.com/en/mobility-report/reports/november-2019

[2] J. Pandey and B. P. Otis, "A sub-100μW mics/ISM band transmitter based on injection-locking and frequency multiplication," *IEEE Journal of Solid-State Circuits*, vol. 46, no. 5, pp. 1049–1058, May 2011.

[3] M. S. Jahan, J. Langford, and J. Holleman, "A low-power FSK/OOK transmitter for 915 MHz ISM band," in *Proc. IEEE Radio Frequency Integrated Circuits Symp. (RFIC)*, May 2015, pp. 163–166.

[4] F. Kuo, M. Babaie, R. Chen, K. Yen, J. Chien, L. Cho, F. Kuo, C. Jou, F. Hsueh, and R. B. Staszewski, "A fully integrated 28nm Bluetooth low-energy transmitter with 36% system efficiency at 3dBm," in *Proc. ESSCIRC Conf. 2015 - 41st European Solid-State Circuits Conf. (ESSCIRC)*, Sep. 2015, pp. 356–359.

[5] M. M. Izad and C. Heng, "A 17pJ/bit 915MHz 8PSk/O-QPSK transmitter for high data rate biomedical applications," in *Proc. IEEE 2012 Custom Integrated Circuits Conf*, Sep. 2012, pp. 1–4.

[6] X. Chen, J. Breiholz, F. B. Yahya, C. J. Lukas, H. Kim, B. H. Calhoun, and D. D. Wentzloff, "Analysis and design of an ultra-low-power Bluetooth low-energy transmitter with ring oscillator-based ADPLL and 4 × frequency edge combiner," *IEEE Journal of Solid-State Circuits*, vol. 54, no. 5, pp. 1339–1350, May 2019.

[7] T. Schumacher, M. Stadelmayer, T. Faseth, and H. Pretl, "A review of ultra-low-power and low-cost transceiver design," in *Proc. Austrochip Workshop Microelectronics (Austrochip)*, Oct. 2017, pp. 29–34.

[8] Y. Tsai, C. Lin, B. Wang, and T. Lin, "A 330-μW 400-MHz BPSK transmitter in 0.18-μm CMOS for biomedical applications," *IEEE Transactions on Circuits and Systems II: Express Briefs*, vol. 63, no. 5, pp. 448–452, May 2016.

[9] Y. Guo, S. Mai, Z. Weng, H. Liu, H. Jiang, and Z. Wang, "A 9.4 pJ/bit 432 MHz 16-QAM/MSK transmitter based on edge-combining power amplifier," in *Proc. IEEE Int. Symp. Circuits and Systems (ISCAS)*, May 2017, pp. 1–4.

[10] C. Lin, Y. Liu, C. Fu, H. Lakdawala, and T. Lin, "An energy-efficient 2.4-GHz PSK/16-QAM transmitter," in *Proc. IEEE Asian Solid State Circuits Conf. (A-SSCC)*, Nov. 2012, pp. 361–364.

[11] R. R. Manikandan, A. Kumar, and B. Amrutur, "A digital frequency multiplication technique for energy efficient transmitters," *IEEE Transactions on Very Large Scale Integration (VLSI) Systems*, vol. 23, no. 4, pp. 781–785, Apr. 2015.

[12] M. Stadelmayer, T. Schumacher, T. Faseth, and H. Pretl, "A 1.2-v 180-nm cmos low-power multi-band ring oscillator based frequency synthesizer for edge-combining transmitters," in *2020 18th IEEE International New Circuits and Systems Conference (NEWCAS)*. IEEE, 2020, pp. 22–25.

[13] W. Kester, "Converting oscillator phase noise to time jitter," *Tutorial MT-008, Analog Devices*, 2009.

978-1-7281-8494-4/20 $31.00 © 2020 IEEE

An N×6M-Path Filter for Low-IF Applications: Review and Modification

Amin Hazrati
Faculty of Electrical Engineering
Shahid Beheshti University
Tehran, Iran
a.hazratimarangaloo@mail.sbu.ac.ir

Ali Jalali
Faculty of Electrical Engineering
Shahid Beheshti University
Tehran, Iran
a_jalali@sbu.ac.ir

Masoud Meghdadi
Faculty of Electrical Engineering
Shahid Beheshti University
Tehran, Iran
m_meghdadi@sbu.ac.ir

Abstract—This paper consists of two main parts. First, a brief review on two band-pass filters based on N-path passive mixers are given, for low-IF application, along with harmonics rejection techniques. This review provides a good insight about the concept of N-path filters, with focus on the ideas necessary for presenting the proposed filter. Then, a new structure is presented for bandpass filters for use in low-IF applications. The filter structure provides a significant noise-figure performance improvement in comparison with similar works. Furthermore, the size of baseband capacitors are significantly decreased, while obtaining a high quality factor, which saves a considerable die area. Designed and simulated in a standard 90-nm CMOS technology, the filter covers the frequency range from 200 MHz to 1.11 GHz by exploiting two digital clock generators. Over the entire frequency range, the analog power consumption is only 21 mW, and the achieved quality factor is higher than 100.

Index Terms—N-path filter, Band-pass filter, Harmonic rejection, low intermediate frequency, N×M-path filter

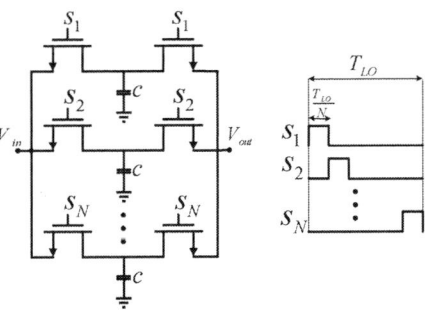

Fig. 1. (a) Traditional architecture of an N-path band-pass filter (b) non-overlapping pulses

I. INTRODUCTION

N-path filters are high-Q filters which can be easily integrated in RF receiver chain, and so they can be fascinating alternatives to the bulky SAW and BAW filters. N-path filters have attracted a lot of attention in recent years, and many researches, including rigorous mathematical analyses and novel designs, have been conducted regarding this kind of filters [1-3]. Fig. 1 shows a very conventional structure of an N-path band-pass filter and the required non-overlapping pulses to drive the switches. The center frequency of the filter is equal to the frequency of the pulses generated by the clock generator. The operation of the filter shown in Fig. 1 can be intuitively explained as follows. The input RF signal experiences down-conversion by the input mixers. The undesired signals are then filtered by the baseband capacitor and the R_{on} of mixers, realizing a baseband low-pass filter. Finally, the desired signal is up-converted to its initial RF frequency by the output mixers. This procedure leads to a high-Q filter for RF applications. However, harmonic selectivity and harmonic fold-back are two drawbacks in this kind of filters, for which some solutions have been presented in the literature [4-8].

As mentioned earlier, assuming the clock frequency is f_{LO}, the sharp band-pass response is obtained at the center frequency of f_{LO}. Thus, this kind of filter can be conveniently

applied in zero-IF applications. However, for use in low-IF applications, a demanding high-resolution digital clock generator circuit would be required to shift the center frequency of the filter by a few MHz or hundreds of kHz, in order to adapt the filter to the desired low IF frequency. In [9], by replacing the baseband low-pass filters with complex baseband filters, the center frequency of the filter can be shifted upward or downward, but the problem of harmonic selectivity around the largest odd harmonics still remains.

This paper is organized as follows: Section II reviews the concept of N×M-path filter and discusses its transfer function. In Section III, the main idea to reject the undesired 3k-th harmonics (k is an integer) will be reviewed, as well as the resulting transfer function and its implementation architecture. The proposed structure will be presented in section IV, along with the simulation results and comparison with some recent works. Finally, section V concludes the paper.

II. N×M-PATH FILTER

In this section, an N×M-path filter and its transfer function are discussed. As mentioned before, N-path filters are well-suited for zero-IF receivers, in which the desired signal and the LO are at the same frequency. In low-IF receivers, and in addition to the LO frequency, another frequency shift in the transfer function is required to place the pass band of the filter

978-1-7281-8494-4/20 $31.00 © 2020 IEEE

2020 Austrochip Workshop on Microelectronics

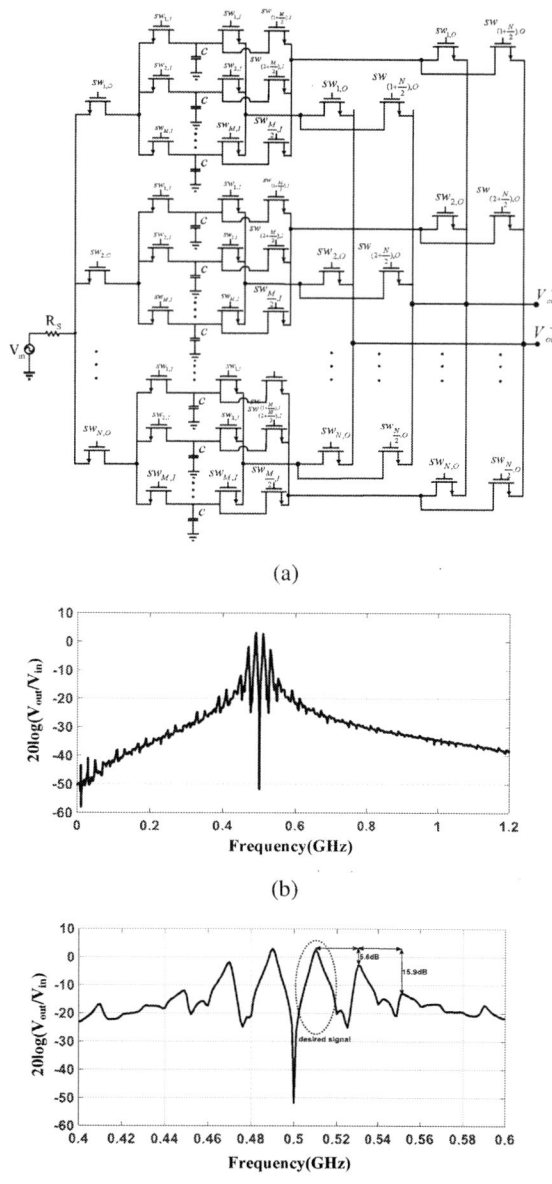

(a)

(b)

(c)

Fig. 2. (a) Structure of a fully differential 4×6-path filter [10] (b), (c) Transfer function of the filter at the desired center frequency of 510 MHz

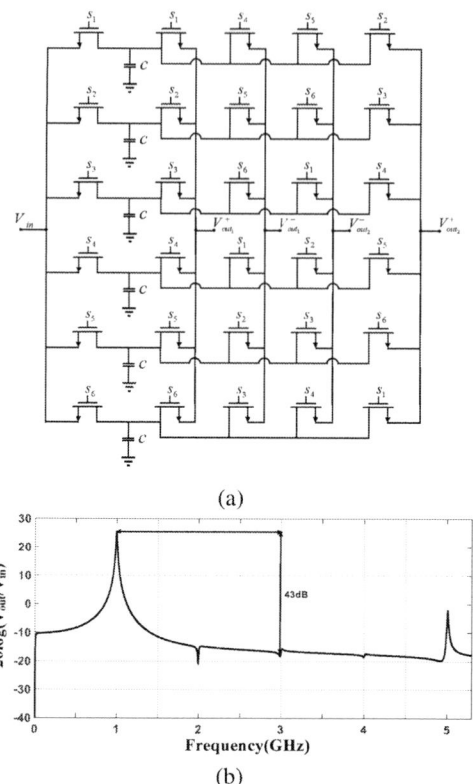

(a)

(b)

Fig. 3. (a) Differential 6-path BPF with 3k-th harmonics rejection capability (b) Transfer function of the harmonic rejection filter [11]

atop the desired signal. An N×M-path filter can be employed for this purpose, as previously introduced in [10]. In order to implement an N×M-path filter, the baseband capacitors in a conventional N-path filter are replaced with conventional M-path filters. The single-ended input to differential output structure is shown in Fig. 2(a).

The switches of the outer, external, N-path filter and the internal M-path filters are driven by voltage non-overlapping pulses with the frequency of $f_{LO,N}$ and $f_{LO,M}$, respectively. It is clear that the input RF signal experiences a two-step down conversion to lower frequencies, and then returns to the initial frequency by a two-step up conversion. So, the achieved pass-band frequency will be at $f_{LO,N}+f_{LO,M}$. Suppose that the structure consists of a 4-path external filter and 6-path internal filters. Fig. 2(b) shows the transfer function of the resulting 4×6-path bandpass filter. The desired signal is around 510 MHz, while the undesired even harmonics around $2kf_{LO,N} \pm hf_{LO,M}$ (where k and h are integers) as well as the surplus harmonics around $f_{LO,N} \pm 2kf_{LO,M}$ (where k is an integer number) are rejected by the differential structure of the external 4-path and the internal 6-path filters. However, many undesired harmonics still remain around the odd harmonics at $f_{LO,N} \pm (2k+1)f_{LO,M}$, because of the harmonic selectivity problem in N-path filters. As it is obvious from Fig. 2(c), the gain difference between the desired signal and the largest surplus harmonic at $f_{LO,N} + 3f_{LO,M}$, (i.e., 530 MHz) is only 5.6 dB. The presence of any blocker at this frequency thus significantly affects the performance of the filter.

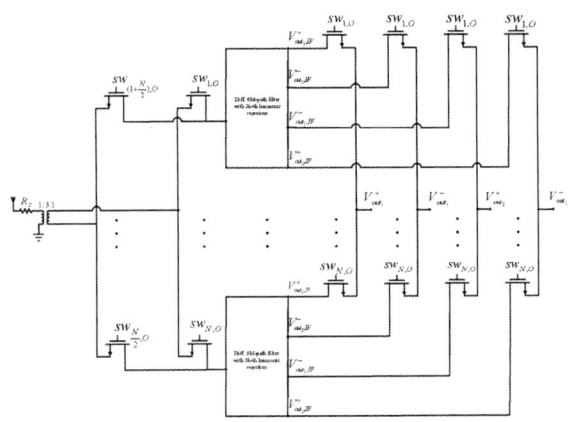

Fig. 4. Proposed structure of the harmonic rejection 4×6-path filter

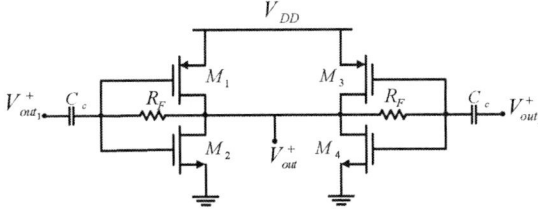

Fig. 5. Wideband resistive feedback LNA [11]

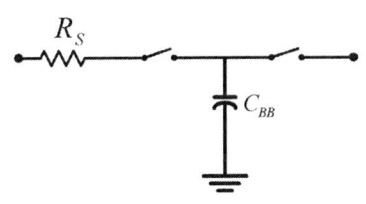

(a)

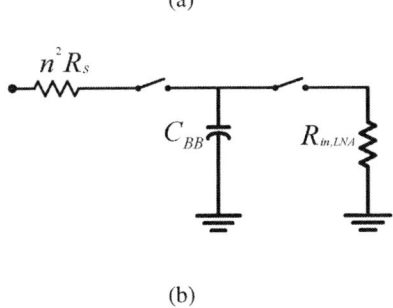

(b)

Fig. 6. (a) Single path of a conventional N-path BPF (b) Simplified single path of the proposed harmonic rejection 4×6-path filter

III. STRUCTURE OF 3K-TH HARMONICS REJECTION N-PATH BPF

As it can be seen from Fig. 2(c), there is a strong demand for an N-path band-pass filter with third harmonic rejection capability, that should be replaced with the internal conventional M-path filters. In [11], a structure for $3k$-th (k is an integer) harmonics rejection has been introduced. Fig. 3(a) shows a differential 6-path band-pass filter which rejects the third and its multiple integers harmonics. By implementing the output mixers with $120°$ phase difference and additional mixers with $180°$ phase differences, respectively, both $3k$-th and even harmonics are rejected. Fig. 3(b) shows the realized transfer function of the mentioned filter, which proves the concept of the applied harmonic rejection. As can be seen, both third and even harmonics experience a substantial suppression.

IV. PROPOSED N×6M-PATH FILTER

As stated in Section I, a number of methods have been introduced to suppress the undesired harmonics in N-path filters. Elaborated in [8], a method for rejecting the odd harmonics is implementing different gain coefficients in the paths of an N-path filter to generate an approximation of a staircase sine wave of the LO. Similarly, in the filter presented in [10], the different gain ratios among the paths of the band-pass filter are generated by exploiting different resistive coefficients, instead of employing different G_m values. As a result, fully passive harmonic rejection 6-path filters were used as the internal filters in the structure of a 4×6-path filter.

From the noise figure point of view, however, and due to the use of resistive coefficients in [10], the noise performance degrades drastically. Furthermore, to achieve a high-Q filter, the total amount of baseband capacitors had to be increased up to 250 pF for each path of the internal 6-path filters, which resulted in a total capacitance of 6 nF, which occupies a very large chip area.

In this paper, by the use of the depicted differential harmonic rejection 6-path filter in Fig. 3(a) as the internal filters, a new architecture has been introduced for this kind of filter, as illustrated in Fig. 4. The input is designed differentially, unlike the filter shown in Fig. 2(a) to avoid the complex implementation required for the output switches of the external 4-path filter. Therefore, an RF balun is utilized at the input to convert the single-ended input to a differential signal. In addition, the employed balun is also used to facilitate the input impedance matching with its ratio of 1 to 3.1, which also contributes to the improvement of the noise figure of the structure by amplifying the input voltage with the factor of 3.1. As a results, a noise figure of only 2.4 dB is obtained spectre RF (pss+pnoise) analysis.

In order to combine the output voltages shown in Fig. 4, two wideband self-biased low noise amplifiers (LNA) are employed, as shown in Fig. 5. Each of the LNAs consumes 10.5 mW analog power form a supply voltage of 1.2 V.

TABLE I
COMPARISON TABLE

	This Work	**[10]**	**[1]**
Number of Paths	4×6	4×6	4
CMOS Tech.(nm)	90	90	65
Freq. Range(GHz)	0.2 to 1.11	N/A	0.1 to 1
Switching Frequency	$f_{LO,N}$=500 MHz $f_{LO,M}$=10 MHz	$f_{LO,N}$=1 GHz $f_{LO,N}$=10 MHz	-
Analog Power (mW)	21	0	0
Digital Power (mW)	1.14 @ 510 MHz	1.213 @ 1.01 GHz	2 to 16
Noise Figure(dB)	2.4 @ 510 MHz	9.1 @ 1.01 GHz	3-5
IB IIP3(dBm)	4.2 @ 210 MHz	15.8 @ 110 MHz	14 to 21
IB OIP3(dBm)	29.2 @ 210 MHz	9.8 @ 110 MHz	14 to 21
Size of Cap.(pF)	50	250	50
B.W(MHz)	3.1 @ 510 MHz	2 @ 110 MHz	35
Quality Factor	100 to 305	55 @ 110 MHz	3-29

Therefore, the overall analog power consumption for the proposed structure is 21 mW.

Another advantage of the RF balun in the input is increasing the source resistance seen by the filter, which decreases the bandwidth of baseband RC low-pass filter, without the need for increasing the capacitance. One may say that each path of the proposed filter sees the input resistance of LNAs as well, which can increase the bandwidth of the filter, accordingly decreases the quality factor. It should be noted that the input resistance of the LNA is equal to 5 kΩ, that its effect on the bandwidth of the filter is negligible. To clarify the point, Fig. 6(a) shows a single path of a conventional N-path bandpass filter. In each transition time, the baseband capacitor sees input resistor, which affect the bandwidth as it depends on both R_S and C_{BB} values. Fig. 6(b) shows the simplified model of each path in the proposed structure. The pole magnitude of the low-pass filter depends on the resistors that the baseband capacitor sees in each transition time. Now, in the proposed circuit, the source resistor is magnified by the factor of n^2, assuming the balun ratio is 1 to n. As we know, the bandwidth of the baseband lowpass filter in Fig. 6(b) is in reverse with the term $R_{in,LNA} \parallel n^2 R_s$. In the proposed circuit, $R_{in,LNA} > 10\ n^2 R_s$, so the effect of input resistance of the LNA on the bandwidth of the filter can be ignored. As a result, a significant decrease in the required capacitor size can be achieved, for the same bandwidth. Note that in our investigation on the bandwidth of the N-path filters we ignored the impact of other phenomena like switch on-resistance. Regarding this issue proper analysis can be found in [1]. Accordingly, in this paper, we use 50 pF baseband capacitors, resulting in a the total capacitance of 1.2 nF compared to the 6 nF used in [10], saving a large chip area, while a higher quality factor is also achieved compared to the presented work in [10]. In Fig. 7 (a) and (b), the Spectre RF (pss+pac) simulated transfer function of the presented filter is shown, around the desired center frequency of 510 MHz. The bandwidth of the filter is 3.1 MHz, and as can be seen, the odd harmonic at 530 MHz is rejected. Due to the high out-of-band rejection, the fifth harmonic at 550 MHz also experiences a very good attenuation.

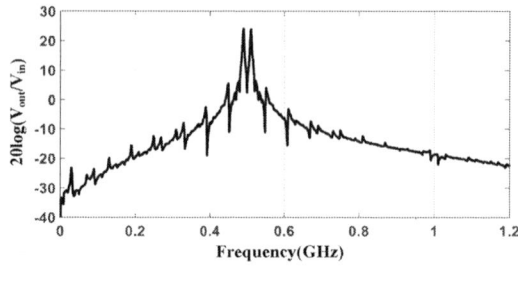

(a)

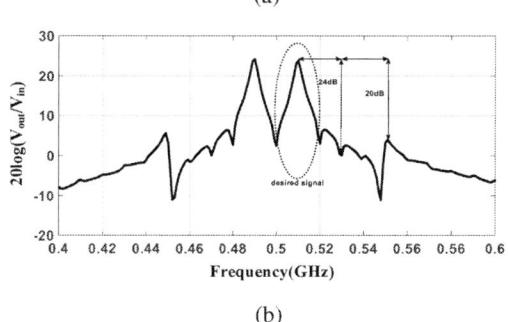

(b)

Fig. 7. (a) , (b) Transfer function of the fully differential 4×6-path filter with 3k-th harmonics rejection capability

It should be noted that to drive the switches of both internal 6-path filters, and external 4-path filter, the clock generator circuitry (which is introduced in [1]) is employed in our filter structure. Indeed, two clock generators are needed to generate the frequencies $f_{LO,N}$ and $f_{LO,M}$. It is well known that the dynamic power consumption is in direct relation to the operating frequency. The most of the digital power is consumed by the clock generator which drives the external 4-path filter switches due to the operation in higher frequencies.

Table 1 compares the proposed work in this paper with two published papers. The results reported in [1] and [10] are based on measurement and schematic simulation, respectively. The presented filter covers a frequency range of 900 MHz, and consumes 21 mW analog power. The other listed filters, due to their fully passive structures, consumes only digital power. The noise figure and area efficiency of the presented filter is highly improved due to the discussed used approaches. Also, a better filtering of the desired signal is achieved because of the higher quality factor compared with the other works, and OIP3 of the presented architecture in this paper is better than reported results in [1] and [10].

V. CONCLUSION

In this paper, the implementation of a fully differential N×6M-path filter was presented which brought the rejection of some surplus harmonics. Tuning range of the proposed filter is varied from 0.2 GHz to 1.11 GHz. The noise figure and total use of capacitor were improved in comparison to its similar work. The filter presents a good behaviour in terms of quality factor and covering a wide range of frequency band. As a result, The proof-of-concept filter is an interesting option for receivers operating with low-IF frequencies.

REFERENCES

[1] A. Ghaffari, E. A. Klumperink, M. C. Soer, and B. Nauta, "Tunable highQ N-path band-pass filters: Modeling and verification," IEEE Journal of Solid-State Circuits, vol. 46, no. 5, pp. 998-1010, 2011.

[2] A. Ghaffari, E. A. Klumperink, and B. Nauta, "Tunable N-path notch filters for blocker suppression: Modeling and verification," IEEE journal of solid-state circuits, vol. 48, no. 6, pp. 1370-1382, 2013.

[3] M. C. Soer, E. A. Klumperink, P.-T. De Boer, F. E. Van Vliet, and B. Nauta, "Unified frequency-domain analysis of switched-series-RC passive mixers and samplers," IEEE transactions on circuits and systems I: regular papers, vol. 57, no. 10, pp. 2618-2631, 2010.

[4] Y. Xu, J. Zhu, and P. R. Kinget, "A Blocker-Tolerant RF Front End With Harmonic-Rejecting N-Path Filter," IEEE Journal of Solid-State Circuits, vol. 53, no. 2, pp. 327-339, 2018.

[5] P. Karami, A. Banaeikashani, B. Behmanesh and S. M. Atarodi, "An N-Path Filter Design Methodology With Harmonic Rejection, Power Reduction, Foldback Elimination, and Spectrum Shaping," in IEEE Transactions on Circuits and Systems I: Regular Papers, doi: 10.1109/TCSI.2020.3009191.

[6] L. Duipmans, R. E. Struiksma, E. A. Klumperink, B. Nauta, and F. E. Van Vliet, "Analysis of the signal transfer and folding in N-path filters with a series inductance," IEEE transactions on circuits and systems I: regular papers, vol. 62, no. 1, pp. 263-272, 2015.

[7] A. Hazrati and A. Jalali, "Signal Folding Improvement in N Path Filters by Programmable Switched-LC Band-Stop Pre-Filtering," 2019 27th Iranian Conference on Electrical Engineering (ICEE), Yazd, Iran, 2019, pp. 228-232, doi: 10.1109/IranianCEE.2019.8786687.

[8] Weldon JA, Narayanaswami RS, Rudell JC, Lin L, Otsuka M, Dedieu S, Tee L, Tsai K-C, Lee C-W, Gray PR. A 1.75-GHz highly integrated narrow-band CMOS transmitter with harmonic-rejection mixers. IEEE J Solid State Circuits. 2001;36:2003-2015.

[9] A. Mirzaei, H. Darabi, and D. Murphy, "Architectural evolution of integrated M-phase high-Q bandpass filters," IEEE Transactions on Circuits and Systems I: Regular Papers, vol. 59, no. 1, pp. 52-65, 2012.

[10] Elmi, M., Poursaadati Zinjanab, A. & Jalali, "A. N × M-path filters: analysis and implementation." Analog Integr Circ Sig Process 96, 543–554 (2018). https://doi.org/10.1007/s10470-018-1170-0

[11] Hazrati, A, Jalali, A, Meghdadi, M, Babaei, B, "A Method for Rejecting 3k-th Harmonics in Bandpass 6N-Path Filters." Int J Circ Theor Appl. 2020; 48: 335– 348. https://doi.org/10.1002/cta.2738

An On-Chip Analog Spectrum Analyzer Based on Miller Frequency Divider

Amin Hazrati Marangalou*$^\diamond$, Santiago M. Sondon$^\square$, Ajinkya Kale$^\diamond$,
Johannes Sturm$^{\diamond\square}$, Michael Gadringer*, Wolfgang Bösch*

*Institute for High Frequency Technology, Graz University of Technology, Graz, Austria
$^\diamond$Frontend Integrated Circuits and Systems, Silicon Austria Labs, Villach, Austria
$^\square$Integrated Systems and Circuits Design, Carinthia University of Applied Science, Villach, Austria
Email: (amin.hazrati, ajinkya.kale, johannes.sturm)@silicon-austria.com, s.sondon@fh-kaernten.at,
(michael.gadringer, wbosch) @tugraz.at

Abstract—In this work, a novel on-chip spectrum analyzer architecture is proposed based on the principle of Miller regenerative frequency dividers. The integrated spectrum analyzer implementation is challenging due to the high Q on-chip filtering requirement and complex down-conversion architecture. The proposed architecture utilizes a frequency divider for second stage down-conversion without the need for external local oscillator signal, thus resulting in a less complex overall architecture. Moreover, thanks to the proposed architecture, the challenge of down-converting wideband modulated input signal in frequency divider architectures is alleviated. The proposed spectrum analyzer architecture is simulated for three test input signals from 2 GHz to 3 GHz with minimum frequency resolution of 1 MHz, input dynamic range of 40 dB and minimum signal level of -41 dBm. The radio frequency input signal is downconverted to 125 MHz at the analog output of the spectrum analyzer with the good Spurious-free dynamic range (SFDR) of 43 dB.

Index Terms—Miller frequency divider, analog spectrum analyzer, local oscillator, dynamic range, chirp signal

I. INTRODUCTION

Spectrum analyzers are ubiquitous measurement instruments utilized for verifying the signals in frequency domain at various stages in a communication system. Advances in the high-data rate communication systems in the recent years has led to need for high speed spectrum analyzers. Spectrum analyzer, like an oscilloscope, can display frequency content of an input signal as well as the waveform in the time domain. The resolution of a spectrum analyzer comes from the high-Q band-pass filter in the path of the input signal. The higher quality factor band-pass filter, the more precise spectrum analyzer will be obtained. Traditional analog spectrum analyzers employ high-Q tunable discrete filters which are swept over the entire frequency range to evaluate the power content in the spectrum [1]. Advance real time spectrum analyzers employ highly complex architectures with many subsystems to provide high accuracy calibrated measurements [2]. Such architectures although very efficient components cannot be completely integrated on to single chip due to performance requirements of the individual subblocks. Integrated spectrum analyzer implementations targeted for built-in-self test (BIST)

of analog circuits have been previously reported [3-5]. Some spectrum analyzers employed wide-band digital intermediate frequency signal processing to replace high Q front-end filters [7]. Some architectures employed switch capacitor filter with external clock signals to provide spectrum sensing capabilities [4-6]. For applications with parallel sensing and processing requirements, architectures employ RF power dividers and SAW filter banks [8]. All the previous implementations of on-chip spectrum sensing, results in complex architectures and reply on external off-chip bulky components for filtering performance requirements.

In this paper, a simplified architecture for on-chip spectrum analyzer is proposed by employing the regenerative frequency divider-based signal down-conversion. Miller regenerative frequency divider is utilized extensively for dividing the clock signals without need for any additional clock signals [8-10]. Frequency dividers are employed in radar target application for linear frequency modulated chirp radar signal down-conversion [11]. However, this led to bandwidth compression after down-conversion. None of the previous implementations employ frequency divider to down-convert a wideband modulated signal. The proposed architecture enables use of frequency divider for down-conversion of wide-band modulated signal without the need for a separate local oscillator signal. This results in simplified architecture for the overall spectrum sensing operation.

The rest of the paper is organized as follows- Section II introduces the concept of regenerative frequency divider and higher order dividing structures. Section III presents the proposed integrated spectrum analyzer architecture, explains the operation, and presents the simulation results for the three test signals along with the discussion. Finally, section IV concludes the paper.

II. REGENERATIVE FREQUENCY DIVIDER

The block diagram of the traditional regenerative frequency divider is shown in Fig. 1. The architecture consists of a mixer followed by a low noise amplifier (LNA), and a tuned band-pass filter at $\frac{1}{2}\omega_{in}$, or a low-pass filter for lower input frequencies. This self-oscillating nonlinear feedback loop, after

2020 Austrochip Workshop on Microelectronics

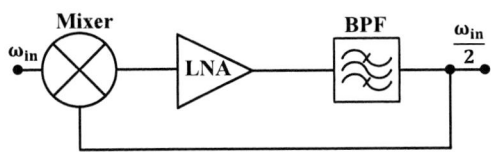

Fig. 1. Conventional architecture of the regenerative frequency divider

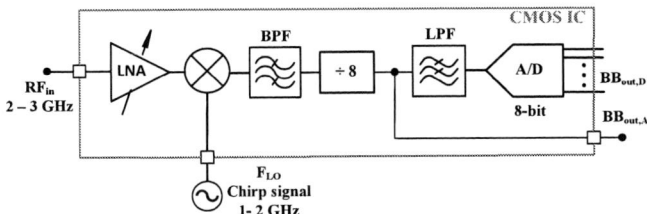

Fig. 3. Proposed architecture of the spectrum analyzer

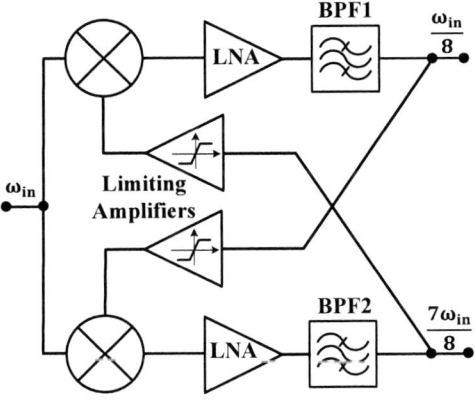

Fig. 2. Regenerative frequency divider-by-eight [13]

settling the divider output signal is at the half of the frequency of the input signal. The mixer downconverts the input signal using the half of the frequency which comes from the tuned band-pass filter. Then, the band-pass filter rejects the undesired upconverted component.

The LNA and the bandpass filter ensure that the mentioned oscillation condition is fulfilled and that only the desired mode of operation of the feedback system is possible.

To support stable oscillations, barkhausen's criterion for stability has to be met. Therefore, a LNA is employed to compensate the losses of loop, and accordingly adjust the loop-gain. The idea of frequency divider can be extended for higher dividing ratio as explained next.

A. Architecture for frequency divider by 8

Divide by 8 regenerative frequency divider block diagram is shown in Fig. 2 [13]. The two bandpass filters BPF1 and BPF2 are tuned at $\frac{1}{8}\, \omega_{in}$ and $\frac{7}{8}\, \omega_{in}$ respectively. So after satisfying the gain and phase conditions, the input signal will be down-converted to $\frac{1}{8}\, \omega_{in}$.

Note that the conventional regenerative frequency dividers suffer from wide range of input signal power at a certain frequency. As we know, the amplitude of the output signal is affected by the input signal amplitude. In other words, whatever the power of of input signal increases at a certain

frequency, the output signal builds up. So to avoid this problem, limiting amplifiers, as shown in Fig. 2, are utilized.

B. Frequency divider for signal down-conversion

Frequency divers have inherent benefit that no external signal is required for clock division. This makes them very attractive solution for down-converting wideband modulated signal specially with higher dividing architectures as presented in the previous section. However, they are designed to a single frequency signal at a time. When a wideband modulated input signal is applied to the frequency divider, due to its non-linear nature, multiple oscillations build-up, degrading the signal to noise ratio (SNR) at the output of the frequency divider and rendering the signal useless. In the next section, the proposed architecture is explained with the solution for this limitation from the frequency divider architecture.

III. PROPOSED ON-CHIP SPECTRUM ANALYZER ARCHITECTURE

The proposed architecture of the on-chip integrated spectrum analyzer is shown in Fig. 3. At the input of the architecture, a variable gain LNA (VGLNA) has been utilized followed by a mixer. The dedicated mixer after VGLNA sweeps the input signal by the use of chirp signal, as an external LO, which varies from 1 GHz to 2 GHz. Importantly, the considerable part of the proposed architecture is elimination of the need for the second LO for further downconversion of the input RF signal that is realized by the use of on-chip frequency divider-by-8. The presented spectrum analyzer covers the input signal power of -41 dBm to -1 dBm. To compensate the loss of the signal chain including mixer and bandpass filter, the gain of the VGLNA varies from 5 dB to 25 dB. Downconverted RF signal at 1 GHz experiences a sharp filtering before second stage of downconversion.

The frequency resolution of a spectrum analyzer is set via the bandwidth of the band-pass filter that is called resolution bandwidth (RBW). So as mentioned above, to sense the input modulated signal in high resolution frequencies, a very high-Q band-pass filter is needed. Such integrable high-Q Band-pass filters are designed in different technologies. In [14], a very narrowband N-path band-pass filter, with the bandwidth of 150 kHz at 1GHz center frequency, was achieved in CMOS technology that its quality factor exceeds 6000. Ultra-narrowband microwave photonic filters are also interesting

978-1-7281-8494-4/20 $31.00 © 2020 IEEE

91

2020 Austrochip Workshop on Microelectronics

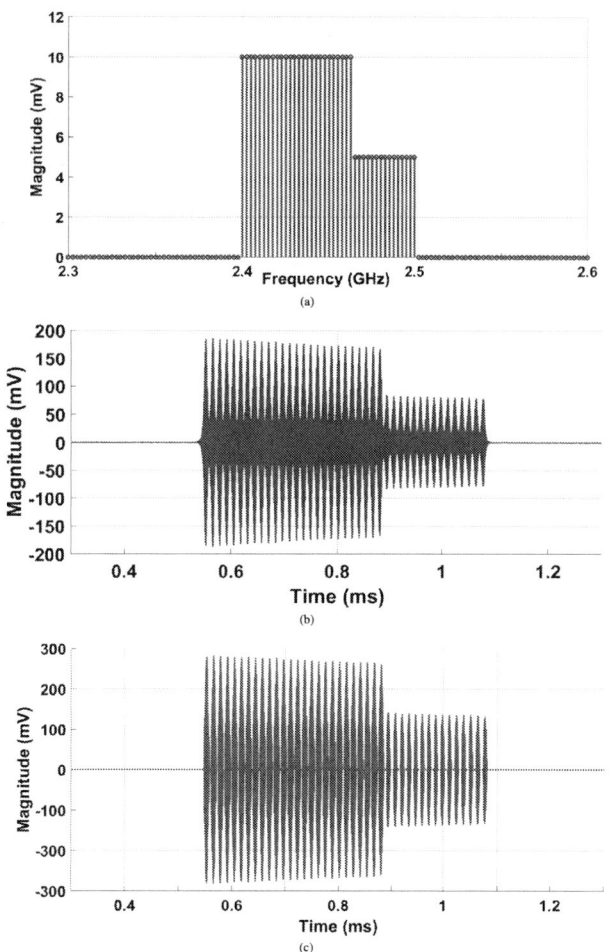

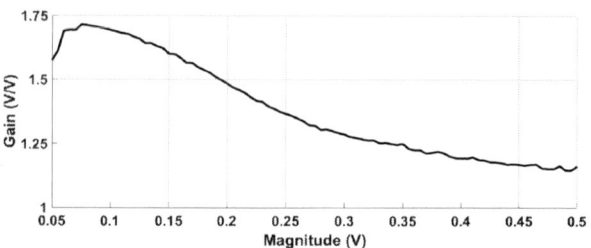

Fig. 5. Gain of the frequency divider-by-8 reference to its input

Fig. 4. (a) Input spectrum with frequency resolution of 2.5 MHz (b) Output of the High-Q band-pass filter at 1 GHz (c) Analog output of the spectrum analyzer at 125 MHz

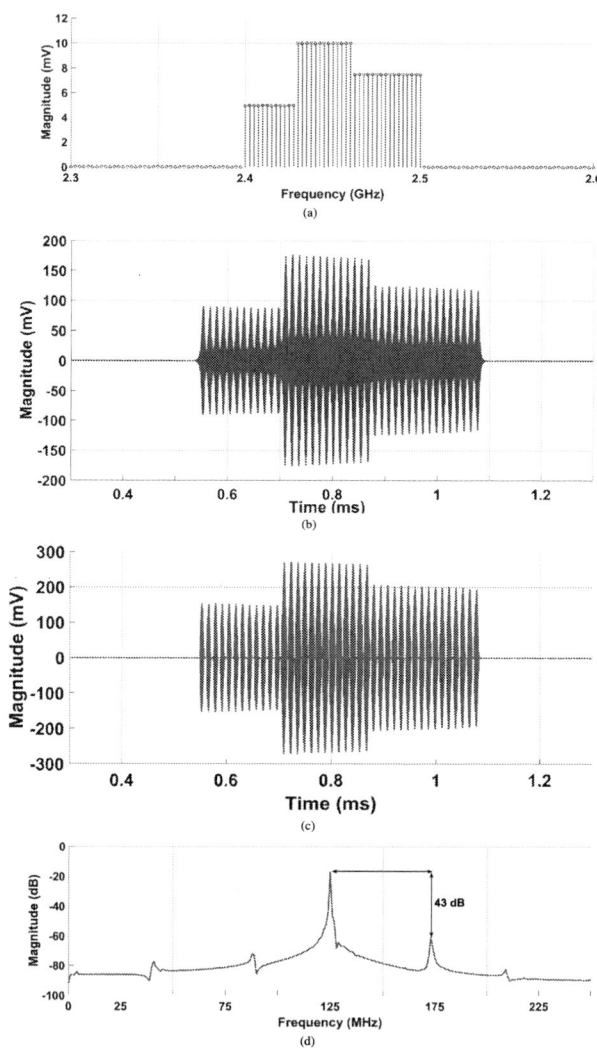

Fig. 6. (a) Input spectrum with frequency resolution of 2.5 MHz (b) Output of the High-Q band-pass filter at 1 GHz (c) Analog output of the spectrum analyzer at 125 MHz (d) output of the spectrum analyzer for input single tone with the power of -36 dBm

options to be used in the architecture of the proposed spectrum analyzers. An RF photonic filter has been designed in [15] that reaches the bandwidth of 100 kHz at RF frequencies. In [16], a tunable microwave photonic filter has been presented that its quality factor is more than 6000. For the proposed architecture, integrated microwave photonics based bands pass filter with center frequency of 1GHz and Q-factor of 6000 is employed.

As mentioned in the previous section, the regenerative frequency dividers are designed to downconvert the single frequency signal. So they are unable to downconvert a wideband modulated signal at a certain time. The most positive point of the presented innovative architecture in this paper is solving the mentioned drawback in RFDs. So downconverting a wideband signal to few MHz or hundreds of kHz is possible now as it will be shown by the simulation results in the next section. After dividing stage, the lowpass filter is used as an antialiasing filter before the ADC. The ADC is running at 300 MHz with 8 bits to support the SFDR at the output of the frequency divider.

IV. PERFORMANCE RESULTS

The proposed system architecture is implemented using Verilog-AMS behavioral models. The performance of the

978-1-7281-8494-4/20 $31.00 © 2020 IEEE

2020 Austrochip Workshop on Microelectronics

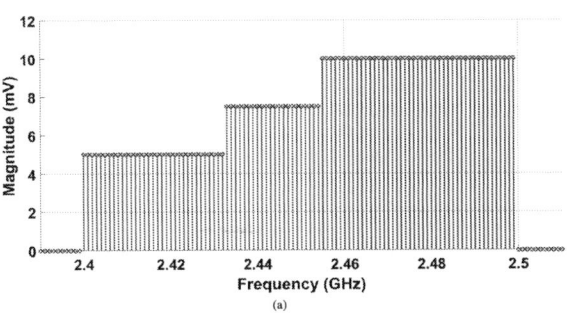

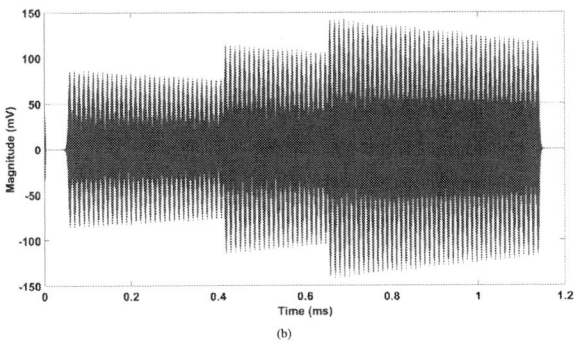

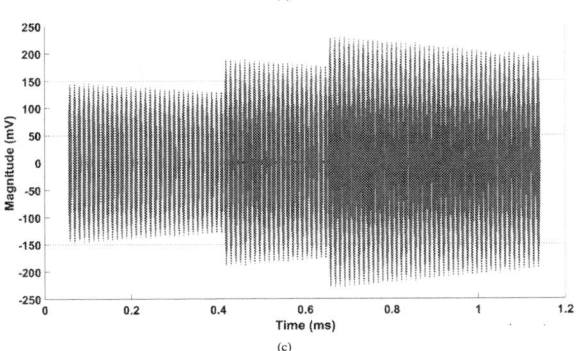

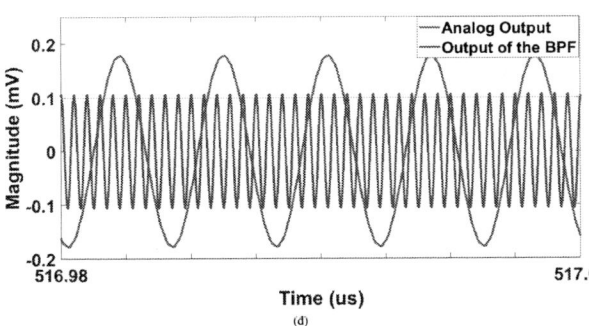

Fig. 7. (a) Input spectrum with frequency resolution of 1 MHz (b) Output of the High-Q band-pass filter at 1 GHz (c) Analog output of the spectrum analyzer at 125 MHz (d) Analog output of the spectrum analyzer and bandpass filter for input single tone with the power of -29.5 dBm at 2.444 GHz

TABLE I
SPECIFICATION TABLE

Parameters	Range
Input Frequency (GHz)	2 to 3
Output Frequency (MHz)	125
LO Frequency (GHz)	1 to 2
Input Power (dBm)	-41 to -1
Input Dynamic Range (dB)	40
Minimum Frequency Resolution (MHz)	1
Analog Output SFDR (dB)	43

to 2.5 GHz with frequency spacing of 2.5 MHz wherein the first 26 tones have 10 mV amplitude and next 15 tones have 5 mV amplitude as seen in Fig. 4(a). The signal after high-Q bandpass filter at 1 GHz is seen in Fig. 4(b). Due to the linearly frequency modulated chirp LO signal utilized in down-conversion mixing, the amplitude information from each of the tones in Fig. 4(a) is translated to time domain signal amplitude in Fig. 4(b). After frequency divider, the down-converted output signal at 125 MHz is seen in Fig. 4(c) where all the tones are resolved. Moreover, the amplitude for the first 26 tones is 270 mV and for the next 15 tones is 120 mV. The gain vs input signal power of the frequency divider by 8 as shown in Fig. 5, shows that the gain of the frequency divider is varying for the input power level. This information can be utilized to compensate the amplitude information at the analog output in digital post processing.

Similarly, for the second test signal multi-tone signal from 2.4 GHz to 2.5 GHz with tone spacing of 2.5 MHz. The first 12 tones have 0.5 mV amplitude followed by the next 13 tones with 10 mV amplitude and the last 16 tones with 7.5 mV of amplitude. The FFT of the second test signal is seen in Fig. 6(a). The output of bandpass filter for the second test signal is seen in Fig. 6(b), where the 1 GHz signal amplitude in time represents the amplitude of individual resolved tones from Fig. 6(a). The output of the frequency divider is shown in Fig. 6(c) where all the tones are resolved, and the amplitudes of the tones match the input signal. Moreover, in Fig. 6(d) the spectrum of analog output (BBout,A) is seen where SFDR of 43dB is achieved.

Finally, the third test multi-tone signal from 2.4 GHz to 2.5 GHz with tone spacing of 1 MHz is utilized. Here, the first 33 tones have the amplitude of 5mV, followed by 22 tones with 7.5 mV amplitude and the last 45 tones of 10 mV amplitude. The FFT of the third multi-tone test signal is shown in Fig. 7(a). These tones are resolved in the Fig 7(b) after the BPF and in Fig 7(c) after the frequency divider down-conversion with the amplitude of 145 mV, 180 mV and 230 mV. The decrease in the amplitude in each subset of tones seen in Fig. 7(b) is attributed to the high-Q filter implemented in the proposed architecture. The frequency divider ratio of 8 can be seen in Fig. 6(d) where for the input tone at 2.444 GHz, the correcting signal at the output of BPF and analog output (BBout,A) is shown. Table 1 also summarizes the performance of the proposed spectrum analyzer.

proposed architecture is verified through simulations for three multi-tone test signals. Matlab based scripts are utilized to generate wideband multi-tone signals with defined frequency resolution to verify the overall architecture.

The first test signal is a multi-tone signal from 2.4 GHz

978-1-7281-8494-4/20 $31.00 © 2020 IEEE

Therefore, based on the simulation results presented in this section, the operation and the performance of the on-chip integrated spectrum analyzer is verified. By utilizing the frequency divider based for 2nd stage down-conversion need for additional LO signal is eliminated leading to less complex overall architecture for the spectrum analyzer. Furthermore, use of chirp signal for the first stage down-conversion enables the frequency divider-based signal down-conversion in the proposed architecture. However, the proposed design relies on post-processing and compensating the gain variation of the frequency divider and the effect of decrease in amplitude due to the high Q filter.

Nevertheless, as highly integrable, less complex architecture the proposed architecture is a next step for on-chip spectrum sensing and built-in self testing for modern communication systems.

V. CONCLUSION

This work proposed an on-chip spectrum analyzer architecture based on the principle of Miller regenerative frequency dividers and utilizing linear frequency modulated chirp signal for down-conversion. Thanks to the photonics based high Q on-chip filter, the high divide ratio frequency divider could be utilized for signal down-conversion with the need for external LO for second down-conversion. The proposed spectrum analyzer operates in the frequency band of 1 GHz with the LO range of 1 GHz to 2 GHz. The performance of the proposed architecture was verified with three multi-tone test signals with the minimum frequency resolution of 1 MHz, input dynamic range of 40 dB and minimum signal level of -41 dBm. The presented architecture in this paper could be the next step for on-chip built in self-test for analog and communication systems in the future.

VI. ACKNOWLEDGMENT

This work has been done in the RF Frontend Lab, jointly supported by FH Kärnten (CUAS) and by Silicon Austria Labs (SAL), owned by the Republic of Austria, the Styrian Business Promotion Agency (SFG), the federal state of Carinthia, the Upper Austrian Research (UAR), and the Austrian Association for the Electric and Electronics Industry (FEEI).

REFERENCES

[1] J. A. Barnes, "A digital equivalent of an analog spectrum analyzer," 1993 IEEE International Frequency Control Symposium, Salt Lake City, UT, USA, 1993, pp. 270-282, doi: 10.1109/FREQ.1993.367406.

[2] F. Ramian, "Implementation of Real-Time Spectrum Analysis, White Paper, 2015. [Online]. Available: https://scdn.rohdeschwarz.com/ur/pws/dl_downloads/dl_application/application_notes/1ef77/1EF77_3e_Real-time _Spectrum_Analysis.pdf

[3] A. P. Jose, K. A. Jenkins and S. K. Reynolds, "On-chip spectrum analyzer for analog built-in self test," 23rd IEEE VLSI Test Symposium (VTS'05), Palm Springs, California, USA, 2005, pp. 131-136, doi: 10.1109/VTS.2005.63.

[4] M. Mendez-Rivera, J. Silva-Martinez and E. Sanchez-Sinencio, "On-chip spectrum analyzer for built-in testing analog ICs," 2002 IEEE International Symposium on Circuits and Systems. Proceedings (Cat. No.02CH37353), Phoenix-Scottsdale, AZ, USA, 2002, pp. V-V, doi: 10.1109/ISCAS.2002.1010640.

[5] G. A. Zimmerman, M. F. Garyantes and M. J. Grimm, "A 640 MHz 32 megachannel real-time polyphase-FFT spectrum analyzer," 1991 Conference Record of the Twenty-Fifth Asilomar Conference on Signals, Systems & Computers, Pacific Grove, CA, USA, 1991, pp. 106-110 vol.1, doi: 10.1109/ACSSC.1991.186423.

[6] Robert A. Witte, Spectrum and network measurements, 2nd Ed., Scitech Publishing, 2014, ISBN: 9781613530146.

[7] D. W. Palmer, R. W. Brocato, L. D. Bacon, G. A. Wouters, W. S. Pickens and G. Loubriel, "Real-time RF spectrum analyzer: Components and system development," 2008 58th Electronic Components and Technology Conference, Lake Buena Vista, FL, 2008, pp. 155-157, doi: 10.1109/ECTC.2008.4549962.

[8] R. L. Miller, "Fractional-Frequency Generators Utilizing Regenerative Modulation," in Proceedings of the IRE, vol. 27, no. 7, pp. 446-457, July 1939, doi: 10.1109/JRPROC.1939.228513.

[9] J. Lee and B. Razavi, "A 40-GHz frequency divider in 0.18-/spl mu/m CMOS technology," in IEEE Journal of Solid-State Circuits, vol. 39, no. 4, pp. 594-601, April 2004, doi: 10.1109/JSSC.2004.825119.

[10] A. Safarian, S. Anand and P. Heydari, "On the Dynamics of Regenerative Frequency Dividers," in IEEE Transactions on Circuits and Systems II: Express Briefs, vol. 53, no. 12, pp. 1413-1417, Dec. 2006, doi: 10.1109/TCSII.2006.886053.

[11] Vorderderfler, M., Gadringer, M.E., Schreiber, H. et al. Frequency dividers in radar target stimulator applications. Elektrotech. Inftech. 135, 344–351 (2018). https://doi.org/10.1007/s00502-018-0626-3.

[12] B. R. Jackson and C. E. Saavedra, "A divide-by-three regenerative frequency divider using a subharmonic mixer," 2011 NORCHIP, Lund, 2011, pp. 1-4, doi: 10.1109/NORCHP.2011.6126726.

[13] O. Momeni, K. Sengupta and H. Hashemi, "Regenerative Frequency Divider with Synchronous Fractional Outputs," 2007 IEEE Radio Frequency Integrated Circuits (RFIC) Symposium, Honolulu, HI, 2007, pp. 717-720, doi: 10.1109/RFIC.2007.380983.

[14] J. W. Park and B. Razavi, "Channel Selection at RF Using Miller Bandpass Filters," in IEEE Journal of Solid-State Circuits, vol. 49, no. 12, pp. 3063-3078, Dec. 2014, doi: 10.1109/JSSC.2014.2362843.

[15] A. Savchenkov, V. Ilchenko, E. Dale, D. Seidel, A. Matsko and L. Maleki, "Agile High-Q RF Photonic Zooming Filter," in IEEE Photonics Technology Letters, vol. 28, no. 1, pp. 43-46, 1 Jan.1, 2016, doi: 10.1109/LPT.2015.2479157.

[16] Chen H. Ultra-wideband microwave photonic filter with a high Q-factor using a semiconductor optical amplifier. Opt Lett. 2017;42(7):1397-1400. doi:10.1364/OL.42.001397.

Author Index

Abbassi F. 37
Arthaber H. 66
Bashir M. 37
Bassi M. 71
Bosch W. 90
Dervić A. 53
Deutschmann B. 47
Eberlein M. 42
Faseth T. 9, 13, 80
Fath P. 9, 13
Gadringer M. 90
Goll B. 66
Hazrati Marangalou A. 85, 90
Hetterle P. 13
Hofbauer M. 62
Hueber G. 37
Huemer F. 21, 29
Jalali A. 85
Jungwirth M. 53
Kale A. 90
Kuttner F. 76
Lang O. 80
Meghdadi M. 85
Mesgari B. 66
Milovancev D. 66

Mortezazadeh Mahani S. 71, 76
Najvirt R. 29
Pichler B. 66
Pretl H. 9, 13, 42, 80
Schmickl S. 9, 13
Schneider-Hornstein K. 62, 66
Schumacher H. 57
Schumacher T. 9, 13, 80
Seebacher D. 71
Sharfuddin Ahmed S. 57
Sondón M. S. 76, 90
Stadelmayer M. 80
Steindl B. 62
Steininger A. 21, 29
Sturm J. 37, 71, 76, 90
Venuturupalli P. 76
Videnovic-Misic M. 37
Vokić N. 66
Weigel R. 13
Zimmermann H. 53, 62, 66
Zupan D. 47

IEEE
445 Hoes Lane
Piscataway, NJ 08854-4141

ISBN 978-1-7281-8494-4